Praise for *Plastic O*

"An ocean hero's call to action."   —Forbes.com

"Fast-paced and electrifying, Moore's story is 'gonzo science' at its best."
                                                                        —*Kirkus Reviews*

"Sobering, impassioned."   —*Publishers Weekly*

"Highly readable, thoughtful, honest, and determined, *Plastic Ocean* is a book with staying power."   —*Booklist*

"In *Plastic Ocean*, readers join Captain Moore on journeys through history, into science labs and to remote parts of the ocean—revealing information both fascinating and incredibly important. A must read for anyone who likes a good adventure and wants to tackle today's pressing environmental problems."   —Annie Leonard, author of *The Story of Stuff*

"The prodigious waste of plastic resources discovered floating in the ocean by Captain Moore brings into stark focus the urgent need to design all our products for complete recovery and reuse within a biological or technical cycle. Only by doing so will our human family become part of a system that replenishes itself on our robust and fragile planet into the infinite future."
                                          —Michael Braungart, author of *Cradle to Cradle*

"Captain Moore set sail across the Pacific and saw what others missed, a plastic plague upon the ocean. This great new book explains that groundbreaking discovery and how our throwaway culture is devastating our sea life."
                        —Laurie David, environmental activist and author of *The Family Dinner*

"Life in our wonderful oceans is stagnating. . . . *This* is what *Plastic Ocean* is really about: a twenty-first-century *Silent Spring*."
              —Captain John W. Koster, Commander, United States Coast Guard Activities Far East

"Captain Charles Moore has turned his childhood love of the sea into a lifelong quest to protect the place he loves. [He is] widely respected for his pioneering work. . . . This book is long overdue. Well done to him and his collaborator Cassandra Phillips!"
                        —Captain Don Walsh, USN (retired), Ph.D., oceanographer and explorer, and
                                                     Honorary President, The Explorers Club

"Charles Moore is the real article. His profound observations and big-picture critique are the sine qua non of ocean protection. He has changed people's life purpose and fulfilled their activist dreams. Without his seminal work and support, we would not have been able to ban plastic bags in 2007 in San Francisco, the first spot in the Western Hemisphere to start pushing back the plastic plague."
                              —Jan Lundberg, independent oil industry analyst, CultureChange.org

"Charlie Moore is the kind of citizen activist who inspires others to stand up for change. *Plastic Ocean* could well become the bible for a whole new generation of environmentalists."
                                               —Ed Begley, Jr., actor and environmentalist

PLASTIC
OCEAN

HOW A SEA CAPTAIN'S CHANCE DISCOVERY

LAUNCHED A DETERMINED QUEST

TO SAVE THE OCEANS

CAPT. CHARLES MOORE

with CASSANDRA PHILLIPS

AVERY

a member of Penguin Group (USA) Inc.

New York

Published by the Penguin Group
Penguin Group (USA) Inc., 375 Hudson Street, New York, New York 10014,
USA • Penguin Group (Canada), 90 Eglinton Avenue East, Suite 700, Toronto,
Ontario M4P 2Y3, Canada (a division of Pearson Penguin Canada Inc.) • Penguin
Books Ltd, 80 Strand, London WC2R 0RL, England • Penguin Ireland,
25 St Stephen's Green, Dublin 2, Ireland (a division of Penguin Books Ltd) • Penguin
Group (Australia), 250 Camberwell Road, Camberwell, Victoria 3124, Australia
(a division of Pearson Australia Group Pty Ltd) • Penguin Books India Pvt Ltd,
11 Community Centre, Panchsheel Park, New Delhi–110 017, India • Penguin Group
(NZ), 67 Apollo Drive, Rosedale, North Shore 0632, New Zealand (a division of
Pearson New Zealand Ltd) • Penguin Books (South Africa) (Pty) Ltd,
24 Sturdee Avenue, Rosebank, Johannesburg 2196, South Africa

Penguin Books Ltd, Registered Offices: 80 Strand, London WC2R 0RL, England

First trade paperback edition 2012
Published simultaneously in Canada

Most Avery books are available at special quantity discounts for bulk purchase for sales
promotions, premiums, fund-raising, and educational needs. Special books or book
excerpts also can be created to fit specific needs. For details, write Penguin Group (USA)
Inc. Special Markets, 375 Hudson Street, New York, NY 10014.

The Library of Congress catalogued the hardcover edition as follows:

Moore, Charles, date.
Plastic ocean : how a sea captain's chance discovery launched a quest to
save the oceans; Capt. Charles Moore with Cassandra Phillips.
p.   cm.
ISBN 978-1-58333-424-9
1. Marine pollution.   2. Plastic scrap—Environmental aspects.
I. Phillips, Cassandra.   II. Title.
GC1085.M67   2011   2011034559
363.738—dc23

ISBN 978-1-58333-501-7

Printed in the United States of America
7 9 10 8 6

BOOK DESIGN BY NICOLE LAROCHE

While the authors have made every effort to provide accurate telephone numbers and
Internet addresses at the time of publication, neither the publisher nor the authors assume
any responsibility for errors, or for changes that occur after publication. Further, the
publisher does not have any control over and does not assume any responsibility for author
or third-party websites or their content.

To the generation, not yet born,

that creates a world where

plastic pollution is unthinkable

CONTENTS

A NOTE TO THE READER

The format of this book is somewhat unconventional, not unlike the format of my life thus far. We have spliced together several stories. One is the story of my evolution as a citizen-scientist who learns how to shape a message and finds himself living it. Another thread can be considered a "tell-all" about plastics, the shape-shifting, inescapable material that seemed like a fun new friend at first but over time has shown its troubling true colors. I only wish we had noticed them sooner. Do not expect strict chronological order. We're dealing here with fragments, in space, place, and time, where boundaries between land and ocean are porous and barely seem to exist, and yet the dots do connect. This is also a love story, because I would not be doing any of what I do were it not for my deep and lifelong love of the ocean.

THE AUTHORS also wish briefly to illuminate the origin of their collaboration, which began on Hawaii's Big Island, in September 2008. The captain had decamped to an upcountry family property following a speaking engagement in Honolulu. Less than two miles away, at an orchid farm, Cassandra had launched Phase II of a USDA-funded investigation of recycled plastics as orchid growth media. She'd noticed in preliminary growth trials that plastics were not, as claimed, inert. Some types stunted, some enhanced growth, and some outright killed (synthetic carpeting). The local library was hosting a Zero Waste meeting. Charlie and Cassandra both showed up—Charlie because he knew the Mainland consultants running it, Cassandra because she hoped to connect with recycled plastic sources. Charlie was curious about the growth trial and stopped by the nursery soon thereafter. Cassandra was reluctant to know more than she'd observed firsthand about the unsettling properties of plastics, but she and her husband, Bob, couldn't help but listen raptly as Charlie described his work and findings. It was Bob who uttered the fateful words: "This could be a book."

CHAPTER 1

A PLASTIC SOUP

THE OCEAN LOOKS like glossy blue cellophane, like a pond on a summer day. The sails are slack, and so are our hopes for a brisk trans-Pacific voyage. We're becalmed in the mid-Pacific "high," and it's not what anyone signed up for. Not Howard Hall, an old family friend, retired prep school dean and math teacher, and veteran deepwater sailor. Not the newlyweds, Howard's son John and his bride, Leslie, who are making this voyage their honeymoon. Not me, captain of ORV (Oceanographic Research Vessel) *Alguita*, a fifty-foot Tasmanian-built catamaran, new enough that we still have a few things to learn about each other.

Nothing is riper with adventure than mapping a voyage, stocking the ship, casting off, pointing the bow toward the open ocean, checking the wind, and scrambling to set the sails. But our overriding concern at this moment, eight days out of port, is getting from point A, Honolulu, to point B, Santa Barbara, before running out of fuel and bonhomie. We've deviated from the standard route due to exceptional conditions. What later turns out to be the most massive El Niño event on record is radiating through the North Pacific. Within a few weeks it will cause torrential rain and flooding in Chile. A marlin will be caught off the coast of Washington, having fled overheated Mexican waters. The trade winds seem to be holding their breath, compelling us to fire up *Alguita*'s engines. It crosses my mind that *The Rime of the Ancient Mariner* starts at a wedding and winds up in the doldrums, like us, "As idle as a painted ship / Upon a painted ocean."

I'm an old salt myself—technically, middle-aged—but this is my first crossing from Hawaii to California, though I've sailed twice from California to the islands. We'd embarked on the standard route, which is to haul well north of Hawaii, then catch westerly trades that sweep you straight across to the California coast. Now the brisk winds that sped our way to the 35th parallel have died down to a whisper. We'd hoped to make it to the 40th before turning right and heading for the West Coast. I check daily weather faxes from the National Oceanographic and Atmospheric Administration and spot evidence of westerlies slightly south of where they would normally be. So we take a gamble and head southeast, into an atmospheric phenomenon called the North Pacific High—also known as the doldrums, or the "horse latitudes," where earlier mariners dumped livestock to lighten their ships and conserve drinking water. It's a more direct route to our destination, but a "shortcut" that could turn out to be long. The wind does pick up . . . for a few hours, but soon we're back on diesel power. Howard and I agree it's highly unlikely this calm will persist much longer. But if it does, we have a problem. And the problem is that we also need fuel for the ship's electrical generator, which runs our onboard desalinization unit and charges batteries for our communication equipment. We're well stocked with food, but our only source of freshwater is the de-sal system. On the slim chance we get into big trouble, we need the ability to radio for help . . .

I begin to notice that this smooth "painted ocean" seems to be—how best to put it?—littered. Here and there, odd bits and flakes speckle the ocean surface. I believe they are mostly made of plastic. It seems odd and improbable. I do not note a first sighting in the ship's log, so I don't know its exact day or hour. My best guess is August 8 or 9, 1997. Nor do I log sightings on successive days, or that I've made a game of it. Days at sea are taken up by responses to changing conditions, dealing with equipment failures, and observing seafaring protocols. This means hourly entries in the ship's log at the helm, visits to the engine rooms, and random trips to the galley for food or a nap in your bunk if you'd pulled watch in the wee hours. My game is this: each time I come out on deck from the bridge, I make a bet with myself that *this time* I will not see another plastic scrap. But I always, always lose. No matter the time of day or how many times a day I look, it's never more than a few minutes before I sight a plastic morsel bobbing by. A bottle here, a bottle cap there, scraps of plastic film, fragments of rope or fishing net, broken-down bits of former things.

This might have seemed "normal," in a dismal way, if we'd been sailing near my home port south of Los Angeles. But we're halfway between Hawaii and California, a thousand miles from land, a place less likely to be littered than the moon, you'd think. Every day for the next several days, as we motor across the eerily still waters of the mid-Pacific doldrums, they are always there: plastic shards, fluttering like lost moths in the surface waters of the deep, remote ocean. I'm bothered, but I'm also distracted by other matters.

Let it be said straight up that what we came upon was *not* a mountain of trash, an island of trash, a raft of trash, or a swirling vortex of trash—all media-concocted embellishments of the truth. It would come to be known as the Great Pacific Garbage Patch, a term that's had great utility but, again, suggests something other than what's out there. It was and is a thin plastic soup, a soup lightly seasoned with plastic flakes, bulked out here and there with "dumplings": buoys, net clumps, floats, crates, and other "macro debris." I was not a latter-day Columbus discovering a plastic continent. I was a seafarer who noticed—at first incredulously, then with greater certainty—that this immense section of the northeastern Pacific Ocean, about halfway between Hawaii and the West Coast, was strewn throughout with buoyant plastic scraps.

I now think of this crossing as both a culmination and a beginning. It turned out other people had seen what I saw, but I hadn't a clue of this. Looking and seeing are two different things, just as thinking something's wrong is a ways away from trying to make a wrong thing right. Who I happened to be in the summer of 1997 was a seafarer with well-developed sensitivities about the natural world. I could not have resisted where they were to take me.

The ocean is the largest habitat on earth, home to more than twice the number of species found on land, with more discovered all the time. The ocean is the planet's womb. Life mysteriously began in the warm broth of a much younger ocean. There it evolved for three billion years before the first creature, spore, or seed found itself on dry land and survived. We're drawn to the ocean. We pay premiums for houses near it, restore our spirits on its shores, and test our courage in its embrace. We coexist with the ocean, but can we really know it? It's like the planet next door, with creatures more alien than those we imagine coming to earth from outer space. The ocean's finned denizens zip through vast tracts of liquid space while we plod around on hard surfaces, needing our inventions to get us where we want to go without straining or taking forever. Now there are so many of "us," nearly seven billion, with all our stuff, lots of stuff, stuff that easily gets away from us. And as I look at these plastic bits bobbing around in the middle of the ocean, it seems the hinge between land and sea has swung a little too wide.

I love the vast ocean for its sameness superimposed on infinite variation. I've been in its thrall my entire life and consider myself a marine mammal. I grew up and still live in a house on the shore of Alamitos Bay, a residential harbor twenty-one miles south of Los Angeles, where waterfront homes are cheek by jowl but most have boat slips and a route to freedom. Summers and after school were filled with swimming, diving, surfing, waterskiing, rowing our dinghy, and sailing with my father in his forty-foot ketch, the *Pink Lady*. Yes, it was pink. A Newporter ketch, whose name and color were not changed when he bought it, a concession to the nautical superstition that it's bad luck to change a ship's name. We—my father, mother, and two younger sisters—sailed the *Pink Lady* to Hawaii in 1961, when I was in my early teens. This was a time when sodas came in returnable glass bottles topped by metal caps. Bic lighters had yet to replace Zippos, people brought purchases home in paper bags, and fishermen used nets made from hemp, manila, or cotton and set them in the ocean on hollow glass balls or wooden floats.

I would have recalled if we'd encountered man-made debris on this crossing because my father would have had a fit. He loved the ocean as much as any man ever did and had a refined sense of matter in its proper place. He worked at my mother's father's company as an industrial chemist, but his natural curiosity took him far afield. For example, garbage dumps held a strange fascination for him. On family vacations, we could expect a side trip to the local dump. We'd all get out of the car and have a good look. I don't recall exactly what he was investigating, but these idiosyncratic forays made their mark. In the 1950s my father began to notice floating trash as he rowed his dory around Alamitos Bay a few times a week. He went to the city and proposed that they contract with him to clean up the bay when he took the boat out. It didn't come to pass. That officialdom dismissed my father's offer of stewardship was formative. Authority has an agenda, but it's not necessarily the right one. I was headed for a more do-it-yourself brand of environmentalism.

I came of age in the sixties and strongly felt the pull of the antiwar movement. I was enrolled at UC San Diego, close to a degree in chemistry and Spanish, when I bolted for Berkeley. We set up an urban commune and got a printing press. We handed out leaflets, thinking we'd usher in a new order that favored people over corporations, peace over war. We questioned assumptions, a habit I've never gotten out of. I decided I needed a trade and joined a cabinetmakers' co-op in Santa Barbara, where I learned woodworking. I moved back to Long Beach and made cabinets for a while. Then I set up a furniture repair shop where we fixed all manner of broken things, including an antique grandfather clock belonging to Vin Scully, the venerable voice of the L.A. Dodgers, and a piano for Gregory Peck, aka Atticus Finch. I ran this shop for twenty-five years. During this time, I started the first commercial organic vegetable garden in Long Beach, got myself a twenty-six-foot sloop, the *Kai Manu*, and became commander of the Long Beach chapter of the United States Power Squadrons, a national organization that offers maritime education. I had a settled, comfortable, vaguely countercultural life with my patient and loyal life partner, Samala Cannon, known as Sam.

But I couldn't help but notice that the coastal community I grew up in was morphing into a place I didn't much like. Imprinted as I was by the marine milieu, it pained me to watch runaway development take away wetlands and favorite surf breaks, pollute estuaries and bays, litter sandy beaches, and send plumes of contaminants into coastal waters. I'd grown up swimming in Alamitos Bay, but by the eighties there were times you wouldn't want to, and you'd think twice about eating a fish caught off the pier. In the early nineties I inherited family money and decided to put it where my mouth had been for thirty years. I chartered the Algalita Marine Research Foundation in 1994, intending to do what I could to help restore coastal waters to their pristine state. At the same time, I founded Long Beach Organic, which turned vacant urban lots into organic community gardens. The name Algalita was an inadvertent neologism. Being fluent in Spanish but no scholar, I assumed the diminutive form of *algae*, *alga*, would be *algalita*, the way Lola becomes Lolita. Besides, I liked the way the word "rolled trippingly on the tongue," as Shakespeare would have put it. I chose this name because the original intent of the foundation was to find a way to reforest California coastal waters with giant kelp, technically known as "macroalgae." This vital coastal habitat was succumbing to pollution, overfishing, and stepped-up El Niño events that were overheating coastal waters. I'd gone through the arduous and costly process of becoming a legally constituted 501(c)(3) nonprofit, and I'd duly paid the California secretary of state to reserve the name Algalita. Once everything was in place, I was gently informed by a professor of macroalgae at the Autonomous University of Baja California in Ensenada that there's no such word as *algalita*. As it turned out, I'd have a second shot at getting it right.

My first taste of environmental activism was with groups lobbying for state laws to protect and restore important wetland habitats in both Californias: Alta and Baja. I'd been an early member of Pro Esteros, a binational group dedicated to preserving Baja California's wetlands, and of the Surfrider Foundation, an unlikely organization formed by surfers whose initial cause— public access to and preservation of surfing beaches—evolved into highly effective international coastal eco-advocacy. I headed its local Blue Water Task Force, which had me recruiting environmentally minded citizens for training in coastal water quality monitoring. Litter and debris were concerns, but our focus at the time was on stealth contaminants: bacterial, nutrient, and chemical pollutants coming from the Los Angeles Basin's watershed. We'd collect samples along shorelines and urban rivers, then submit them for analysis at state-certified labs. Later on, we'd process them in our own lab. My impression during this time, in the early nineties, was that the pollution problem was getting away from us. Mankind's "urban runoff" seemed to be escalating beyond what we were able to assess, and going farther out to sea. Sailing offshore in my sloop the *Kai Manu*, I was starting to notice solid trash floating in the open ocean. Before, you'd see trash stranded in the wetlands and harbors, but not out there. I wanted to do some sample-taking in these waters, but the *Kai Manu* was not the ideal research vessel.

So I began exploring options. A proper oceanographic research ship would run in the many millions, and it wouldn't afford the economies and pleasures of an agile sailing ship. It seemed to me the ideal craft to carry out the foundation's work, and possibly earn its keep as a charter, would be a catamaran.

Catamarans had been around Alamitos Bay since the maker of my first surfboard, Hobie Alter, invented the Hobie Cat—a spiffy little sport catamaran—in 1969. I'd noticed how they seemed to fly across the water. The twin hulls confer peerless stability, and if you happened to run aground, your cat would sit upright, unlike a monohull vessel, which would pitch over calamitously on its side. (I was to learn how great a virtue this is.) I liked how cats can scoot along even in a light breeze, because they're not weighed down by ballast. It's a pleasing, fuel-saving efficiency, and a smart, ancient technology known well to the early South Pacific Polynesians who piloted giant catamarans the five thousand miles to Hawaii's shores. I envisioned a boat that would carry me from California to the Hawaiian Islands, a voyage half this distance. Also, just being practical, I needed a vessel with shallow draft to monitor the bays and estuaries of Baja California. So a catamaran it would be: a vessel to test the waters, literally, and take the new foundation, Algalita, to the proverbial next level.

Thus the quest for *Alguita* began, for I'd already decided to name this ship the *correct* diminutive form of *alga*. And it wasn't always smooth sailing.

First, I needed to find a model for *Alguita*. I pored over catalogs and zeroed in on the pioneering Australian catamaran designer Lock Crowther. His line included a vessel with almost every feature I sought. But only one of this particular design had been built, and it was owned by an Australian. In the spirit of due diligence, I contacted the owner, who graciously invited me to "test-drive" the *Sunbird* on a visit to Cairns, Australia. After our spin, I realized certain modifications would be needed to align this fifty-foot aluminum-hulled craft with my vision for a versatile, modern research vessel.

The next step was to find a boatbuilder. I put the project out for bid and picked the lowest, from the firm of Richardson Devine Marine Constructions in Hobart, Tasmania, recommended by the husband of an old Long Beach friend who'd married a "Tassie." Contracts signed, I flew to Tasmania to confer with the team. I wanted to raise the cabin's roof and enlarge its windows to alleviate the cave-like feeling of the main saloon. I also required reverse transom swim steps for easy diver access to the water and wanted the mast positioned closer to the bow. The mast position would allow for a larger mainsail, meaning greater speed and tacking ability, but it also had to be tabernacled (hinged), so I could drop it down and pass under the twelve-foot bridge at our home port of Alamitos Bay.

Alguita was christened with two bottles of champagne—one for each hull—nine months later, in November 1995. I stood back and admired her. She was a trim, well-purposed vessel. Each hull had three bunks and a "head." The galley lay between, on deck level, and the dinette could be made up as a double bed. She had a small wet laboratory, and a wide bow and stern suitable for sample-taking maneuvers involving a small team. Her official name was Oceanographic Research Vessel *Alguita*. Welded onto her aluminum hull was her U.S. documentation number: 1037108.

As pleased as I was with *Alguita*, several early misadventures left me with the uneasy feeling that she might have been born under a bad sign. We put her right to work. The first job was to ferry a team of Tasmanian geologists to remote Black Pyramid Island, which happened to lie in a stretch of notoriously rough seas called the Roaring Forties—after the 40th parallel, where it lies. The geologists needed to reach nearly inaccessible cliffside sites to get the first geologic samples ever taken at this rocky twenty-five-acre islet. We dropped anchor as close to shore as could be safely done and used the dinghy to ferry the research team to and from the inhospitable formation. Leaving Tasmania, we provided a floating platform for University of Sydney scientists studying a marine dead zone at the eastern Australian port of Newcastle, the regrettable result of years of poorly regulated coal loading. Then back to Sydney for a water quality sampling mission with the Aussie Surfrider Foundation along Australia's eastern seaboard. We conducted our first survey of "anthropogenic debris"—marine trash—using a checklist developed for the project. Still, as far as we were concerned, the main event was monitoring for bacterial and nutrient contamination—especially the former, which had led to a rash of beach closings.

It was on this mission that *Alguita* had her first brush with disaster. We were staying over in the harbor at Gladstone, a little coastal town protected by a ring of treacherous reefs. We'd been safely guided out of the harbor by a Surfrider rep from a nearby coastal town. After a day's visit, we dropped him off and headed back to Gladstone for another night's mooring. It turned out we lacked adequate small-scale charts showing the lay of the reefs. In the dimming twilight, trying to find the shortcut we'd taken earlier, I drove *Alguita* hard aground on a semi-submerged formation called Seal Rocks. All but two of her ten watertight compartments were punctured. But, true to form, she was still upright! She floated off the rocks with the rising tide, but there was no way she'd make it to port without foundering. The all-volunteer Gladstone air/sea rescue team efficiently delivered a gasoline-powered pump for bailing water and provided an all-night tow back to Gladstone harbor. It was morning by the time the team parked her at high tide on a city beach, knowing she'd be high and dry for damage assessment when the tide ebbed. In sailor parlance, she was careened. The local Red Cross arrived with sandwiches. We were deeply grateful for them all. *Alguita* spent nearly as much time in dry dock for repairs as it took to build her. But she emerged in better shape than ever, with a thicker aluminum hull and bulbous bows of polyurethane foam. Each bow had been equipped with polished agate "eyes" from Mount Hay Gemstone Park, gifts from the local lapidary club where I killed time in the evenings. She also sported new cabinets, improved electronics, rebuilt engines and dive compressor, and upgraded electrical wiring.

In the fall of 1996, a year after her champagne christening, *Alguita* set sail from Gladstone, this time on her "delivery voyage" to her permanent home in California. We'd chartered her for monitoring work en route, with stops in Fiji and American Samoa. We logged a sighting of a foamed polystyrene cup as we neared American Samoa, where we stayed a few days, helping to clean up a trash-choked stream and replant a mangrove forest.

Our next leg, which had us departing shortly before Thanksgiving, would take us to a layover in Hawaii. But *Alguita* had another surprise in store. As we approached the equator, a line of squalls appeared on the radar. We swiftly moved to reduce sail by "roller reefing" the jibsail, meaning we partially furled the large front sail called the Genoa jib. Large sails in big winds spell disaster. But the wind shifting in advance of the first squall caught the remaining scrap of open jib and set it to flapping violently. With a piercing crack, the top three-fourths of the aluminum mast snapped and crashed down to the port side of the boat, smashing the railing just inches away from Jeffrey Earle, one of my cocaptains. He'd ducked down just in the nick of time. Though badly shaken, we managed to lash the broken mast and rigging to the port side of *Alguita* and motor back to Pago Pago, American Samoa's capital. We contacted boatbuilders in Australia, but none had a replacement mast. Plan B involved making arrangements for *Alguita* to be shipped home.

Pago Pago had become the center for tuna canning for the U.S. market, being closer to the catch and offering comparatively cheap labor and lax regulations. (Pumping cannery waste into the coral-studded harbor was routine.) We secured space on a small container ship, the *Vai Mama*, and had *Alguita* winched to her deck, all twenty-five tons of her. She sat on a bed of plywood and tractor tires, perched on containers filled with canned tuna fish. Her first trip home was not to be triumphal but oddly fitting for a type of boat nicknamed a "cat."

Back in Long Beach, we found a rigger who designed, built, and installed a new aluminum mast twice as strong as the first. He redesigned and improved the rigging system as well. Once again *Alguita* proved her resilience, emerging from this latest mishap stronger than ever. I decided there'd be no better way to test her new mast than to race her, and it happened that an opportunity loomed before us: the Transpacific Yacht Race, a classic open ocean race known as the Transpac. The first race was run in 1906, when three yachts set out from San Pedro Lighthouse bound for Diamond Head, a distance of 2,225 nautical miles. Now an average of seventy boats race in six divisions. In 1997 a catamaran division was added. Only four cats were entered. One was an owner-built craft that capsized soon out of the gate. Another was the sixty-foot trimaran *Lakota*, then holder of the Transpac's speed record. This super-sleek vessel was owned and captained by ill-fated businessman-adventurer Steve Fossett. The third was a formidable eighty-six-foot racing cat, *Explorer*, piloted by Bruno Peyron, a French yachtsman who'd already broken the world speed record for circumnavigating the globe in this very boat.

With her twin diesel engines and built-in research accoutrements, *Alguita* was no racehorse. We'd lightened her up for the race by removing the trawl winch cable, thereby trimming several hundred pounds, but she was still no match for these high-tech Seabiscuits. Her thirteen-day crossing put *Alguita* in the rear of the pack overall, but we earned a third-place trophy in our division of three, and bragging rights. Peyron, the Frenchman, shattered Fossett's Transpac record with an incredible time of five days, nine hours—covering more than four hundred miles per day—shorter by several days than the fastest monohull. *Alguita* had a respectable top speed of 20.5 knots, running downwind in the aftermath of Hurricane Dolores. And unlike five other boats in the race, she held on to her mast. We were well pleased.

El Niños interest me greatly, and there were early signs the event of 1997 was going to be a doozy. El Niño is a cyclical warming trend in equatorial waters that tends to spawn hurricanes, among other disruptions—like no wind at all. With this in mind, we'd prepared carefully for the return voyage. On the Big Island, where I laid over to visit family, we managed to score four sheets of plywood from the mothballed set of the problem-plagued disaster film *Waterworld*, which was shot here over many months in 1995. The plywood would come in handy for boarding up *Alguita*'s windows in case of high winds and breaking seas. It turned out they would be handy for a very different use. *Alguita* is equipped with dual diesel engines and a good-sized fuel storage bladder. According to our original scheme, we'd be able to sail most of the way back on the standard northerly route, alternating engines if the winds slackened.

It's somewhat fascinating to find yourself in the middle of the North Pacific High, an area typically avoided by seafarers. The high is so called because it's an area of high atmospheric pressure. The denser air here has a heating and drying effect, creating what could be called an oceanic desert. In fact, it lies between the same latitudes as the Northern Hemisphere's great deserts in the Southwest, Mexico, and Asia—also high-pressure areas. Even fish tend to avoid the sluggish waters of this oceanic zone. Which is not to say they're biologically bereft. Food chain anchors are here, including the very basis of the marine food web, phytoplankton—tiny marine plants that photosynthesize throughout the world's oceans and generate half the earth's oxygen supply. One step up are the tiny creatures collectively known as zooplankton. These include an exotic array of small gelatinous, filter-feeding organisms that go by names like salps, siphonophores, tunicates, and "jellies." Sea turtles and tuna venture in to dine on these creatures and on small luminous night-feeding fish called myctophids, or lanternfish. Of these organisms I would learn much more.

With me manning the galley, we eat well. It's the doldrums, so there's plenty of time for cooking and eating and swimming and reading. I still don't log my visuals of little plastic floaters, although I sketch out a few rough calculations on a notepad. Howard spots a tantalizing glass Japanese fishing float but it's just beyond our reach. Day 16 is a big one. I hook a hundred-pound big-eye tuna—ahi, the coveted sashimi fish—at eight a.m., when we're making about 6.5 knots. Landing it entails a fierce struggle that has me running up and down the length of *Alguita* as I reel it in. It's our protein ration for the rest of the trip. Ahi day is a red-letter day for me in another respect. At 0530 I pop the spinnaker—a difficult sail to raise—single-handedly for the first time. But after a mere five hours' deployment, in mild 12- to 15-knot winds, our spinnaker halyard—the nylon line that raises this useful sail—snaps. We can't afford not to take advantage of this welcome breeze. I have no choice but to get out the bosun's chair and let the crew winch me to the top of the sixty-five-foot mast to fix it. It's no accident I'm a seafarer, not a mountaineer.

We finally run out of fuel in the port tank at one a.m. on day 18. I'm able to transfer fuel from the starboard tank to the port tank, which is where the all-important generator is housed. Around this time, Howard Hall says we need to go into resource conservation mode. Howard knows whereof he speaks. He and his wife sailed a small sloop from California to Hawaii in the days before faxed weather maps and instant communications. En route, a hurricane overtook them, but they came through un-scathed, not even realizing what they'd just survived. So now we wash dishes in salt water and stop taking showers. Lisa, the bride, makes a log entry that reads "The wind is excruciatingly light." On day 19, another log entry says: "No movement at all!" The charts tell us we've entered coastal waters, where the chilly, strong southerly California Current should be kicking up stiff breezes. But it seems the El Niño has managed to subdue even these reliable wind patterns. We cut our *Waterworld* plywood to make a transom on the stern swim step and attach a small gasoline-powered outboard motor borrowed from *Alguita*'s skiff. This takes us at one knot per hour nearly as far as Santa Cruz Island, northwest of Santa Barbara. We can almost see the California coastline. But we can't get there. Luck finally breaks our way when I get on the radio and reach another Transpac racer, the *Salsipuedes*, docked at Santa Barbara. They bring us two five-gallon jerry cans of diesel. For safety's sake, they toss them in the water near *Alguita*. I swim out and bring them in like a fuel-can lifeguard. This is day 20. With our ten gallons of diesel, we make it to Avalon, Catalina Island. The starboard engine is faltering, making it unsafe to maneuver through the lines of moored vessels to the fuel dock. We need to drop anchor and row the dinghy to the fueling station, refill the jerry cans, and row back out to *Alguita*. We finally reach Long Beach and Alamitos Bay that afternoon.

Back on terra firma, I can't stop thinking about all those miles and days of plastic. Over the decades we'd gotten used to the sight of trash on the beach, by roadsides, and in riverbeds, of shopping bags fluttering on fences and branches, near-weightless foamed polystyrene cups skittering on the breeze, cigarette butts and bottle caps everywhere. . . . And for a while they didn't seem terribly malign. More like annoying evidence of someone's carelessness, or of gusty weather. But something seemed very wrong about this plastic trash in the mid-Pacific. Of all the places on earth, this one should have been exempt. I probably knew then that plastic is slow to decompose but would not have realized it doesn't biodegrade in any meaningful time frame. I had yet to learn that man-made polymers—hydrocarbons bonded by heat and chemical reactions—were strong, persistent molecules. A plastic object decays into fragments, then into nanoparticles that might go on polluting for centuries. What might it mean for living organisms encountering these immortal motes? I was well aware that stray plastics pose ingestion and entanglement risks for marine wildlife in coastal areas. Years before, I'd come upon and rescued a brown pelican that I found struggling helplessly near the Long Beach breakwater. It had been pierced and trussed by a hooked length of monofilament fishing line. But I began to wonder if these errant mid-ocean flecks might be interfering with natural processes at the base of the food chain. It would not have occurred to me at this time, because the science was so new, that these plastic shards could also be toxic.

I searched for the notepad on which, during the voyage, I'd scribbled rough calculations, based on estimates of plastic bits visible from the deck. I saw plastic litter on seven successive days, over about one thousand nautical miles. I figured this soup might cover an area as big as a circle with a thousand-mile diameter. Based on an estimate of half a pound per one hundred square meters, this part of the ocean might contain as much plastic as the two-year total of everything dumped at the United States' largest landfill at the time, Puente Hills, serving the greater Los Angeles Basin: 6.7 million tons.

Our premature shortcut had ended on a note of drama after nearly three long weeks at sea. Lisa, Howard's new daughter-in-law had put on ten pounds, to her great distress. I caught a spectacular fish and had a white-knuckle ride to the top of the mast to fix the broken halyard. And I had seen an ocean strewn with plastic confetti. Back home, I would tell the stories of the fuel shortage, the ahi, and the trip up the mast. But those errant bits and pieces of plastic, adrift where they least belonged—in the most remote part of the largest ocean on the planet—wouldn't leave me alone. The questions kept mounting. Where had they come from? Land? Ships? Fishermen? Despite an international law prohibiting dumping of plastic garbage at sea? Had it come from Asia, the U.S. West Coast? Somewhere else? Where would these wayward morsels wind up? Inside innocent marine creatures? On already polluted beaches, or the ocean floor? Would they hang in eternal suspension in the ocean's surface waters? Was this plastic soup a fluke, something temporary instigated by the massive El Niño? Would it be found in other oceans? Could these rogue plastics be causing ecological distress? I needed to know.

With support from Algalita's directors, I start planning a return voyage to the North Pacific High to verify and quantify the presence of plastics there. I didn't know at the time that this mission would take my life in an entirely new direction. Or that I'd gain a surprising amount of notoriety as the "discoverer" of the Great Pacific Garbage Patch — a term I've never really cottoned to, but appreciate for its utility as a "brand." The patch is not a patch at all, but an immense area larger than all the pumpkin patches on earth. This voyage will acquaint me with new things: scientific controversy, skeptical officialdom, and media glare.

Now I can appreciate the irony. The funds I used to create Algalita and build *Alguita* came from a trust created by my grandfather, Will J. Reid, president of Hancock Oil Company. Petroleum and natural gas provide raw materials for the synthetic polymers known as plastics. And I think of my father, engaged in and extrapolating from his observations at local dumps. Or gamely offering to clean up Alamitos Bay. And it seems somehow inevitable that I would find myself going back to that polluted place, wanting to unravel the mysteries of its synthetic contents. But also to seek redress for the oceans.

CHAPTER 2

SYNTHETIC EVOLUTION

ON A FALL DAY, I have an out-of-town visitor with a curiosity about plastics and my work. Our journey that day bears recounting, as it seems to encapsulate much about the origins and fate of plastic debris. Here we pause my personal narrative and take up the story of plastic itself.

It starts at *Alguita*'s slip, where, to my unpracticed visitor's amazement, I spot a tiny nurdle bobbing near the vessel's hull. Since they amount to about 10 percent of beached plastic fragments, I am able to find one or several in a minute or two for any interested person. I pluck and pocket it before we hop in my Prius and set out. A few miles away, on a tree-lined side street, we pull up to my old woodworking and upholstery shop—now owned by a former employee and well-stocked with polyurethane foam for cushions. A foam cup lies in the gutter and a polyfilm bag dangles from a nearby garbage can. Introductions made and tour given, we motor on to the organic community

gardens I spearheaded more than fifteen years ago in the aftermath of downtown riots. Both lie behind locked gates, but the
crops look well tended and delicious. We're disappointed to see,
at both sites, plastic bags and foam plates and cups blown up
against the fences like ersatz autumn leaves. We then hum up
the highway that crosses the Los Cerritos Channel and wave to
my friend Lenny Arkinstall, hired by the city of Long Beach to
manage seaward-drifting debris. He stands on the riverbank,
near a power shovel, preparing to scoop up the catch of the
day—several thousand square feet of bobbing bottles, jugs, bags,
wrappers, cups, straws, balloons, and, no doubt, shoes, sports
equipment, and other urban detritus—stopped in its tracks by a
net run across the intake to the largest power plant complex west
of the Mississippi.

We pull back onto the highway behind a garbage truck. As if
on cue, a plastic shopping bag whips up and out of the top of the
collection bin and catches the coastal breeze. Our tour is starting
to seem like a video game about chasing down plastic trash. On-
ward we cruise, rubbernecking a couple of plastic-injection and
roto-molding operations, and a "bag blowing" plant. The mass
production of plastic pellets as feedstock for consumer and industrial plastic products is highly centralized. But those end-use
items are often fabricated in smaller shops, called "converters,"
which number about 13,500 domestically, down from 16,000 in
2001, according to the U.S. Bureau of Labor Statistics. We park
in front of a plant where I'd staged an intervention a few years
back. Fluffy plastic shreds and roly-poly nurdles line the edges of
the parking area. The shreds are shavings from the finishing process, the nurdles wasted feedstock, both evidence of bad practices.
A manager comes to the edge of a loading dock and stares us down
like a Marine Corps sergeant. I genially bid him give our best to

the owner, whom I'd had dealings with on matters unrelated to plastic for many years.

We next head over to the Algalita office, where science and administrative staff are wrapping up a meeting. Holly Gray talks about her seabird study. She's examined the stomach contents of seabird bycatch: birds hooked and drowned by fishermen's baited longlines. Christiana Boerger is likewise compiling data from a fish-gut study. Both are investigating ingestion, quantifying how much plastic debris is consumed by marine creatures. We walk across the street to the mouth of the San Gabriel River. The river flows into the sea about twelve feet below street level. Down here near the breakwater, huge granite rocks called riprap buttress the riversides. Following a hundred-year rainstorm in 2004, floodwaters from upstream left a massive load of nurdles on this beach. Years later, it still takes only a few minutes to fill our pockets. Then on to the Sea Lab, where able technician Ann Zellers processes samples from our most recent voyage into the gyre.

Like explorers chugging upstream, we continue on, navigating northward in the Prius, toward the mighty petro El Dorado rivaled in the U.S. only by a few Gulf Coast ports. As much as I claim to be a product of my marine environment, I'm also a lifelong denizen of the West Coast, ground zero for much that doesn't belong in the oceans. Long Beach, my town, neighbors the Port of Los Angeles. Both are nestled behind five miles of breakwater stretching in three sections from the Palos Verdes Peninsula south to the San Gabriel River. In my lifetime, they became the top two container shipping ports in the United States. Combined, they would rank number five in the world, though each by itself would just squeak into the top twenty, distantly trailing mega-ports in China, South Korea, and Europe. To drive north across the Vincent Thomas Bridge, spanning the narrows that divides the two ports, is to gaze down on a city-sized swath of shipping containers, stacked and quietly waiting a call back to service on rails, trailer rigs, and high seas.

It's axiomatic: where there is oil there is plastic. By the 1920s, California was the top oil-producing region in the United States. At the time, car ownership and the petrochemical industry were both gestational. Output outstripped demand, so here, Southern California, is where the world's earliest oil tankers steamed into port empty and left loaded with crude. But those days of export are long gone. Now all the major oil companies have a presence here, and their tankers line up along the coast like jets at LAX, waiting to deliver their carboniferous loads. And a plastics-making infrastructure inevitably sprang up among the refineries.

We pull up to a hangar-sized "nurdle depot," where polypropylene pellets made at a nearby ARCO plant are processed for shipping. A few years earlier I'd given a well-received talk to employees in the plant's central control room about the problems caused by plastic debris, especially bottle caps made from their feedstock. The depot is conveniently located on a rail line. Sure enough, we see several container cars parked alongside the plant. Long Slinky-style vacuum tubes connect them to pellet bins inside. In the old days, staff would have carelessly detached these tubes from the boxcars. The result was ever-growing conical heaps of nurdles along the tracks. Years of spillage were economically insignificant to headquarters but significantly polluting to nearby shores. I was gratified on this visit to see collection pans positioned beneath the hose couplings. Judging by how full they were this day, they'd foiled the oceanward odyssey of tens of millions of nurdles, at least, in the past several years. This plant got its act together back in the 1990s when the EPA began cracking down on plastic pellet pollution. But I've done remediation work at other plants where personnel were clueless about basic pollution laws. We've videotaped places where you couldn't see the ground for the nurdles, and panned the same shot to storm drains—gaping portals to the Pacific.

"But how are plastics actually made?" asks my guest, who'd read about molecules being cracked and polymerized and had no idea what that really meant. We head over tracks and roads worn rough by endless convoys of semitrucks, toward a fantastic skyline backlit by the afternoon sun, stretching north to south as far as the eye can see. With its gleaming domes, minaret-like towers, and fretwork of pipes and raised walkways, the refineries seem a fever dream of a forbidden city. In fact, they are the beating heart of the American way of life, and a neonatal unit for the plastics industry.

We pull up to a security gate and an armed uniformed guard emerges from a nearby trailer. So we park a ways away and watch, and I recall what my father—married to an oilman's daughter, my mother—had told me about cracking molecules. It occurs in the towers technically known as distillation. There, individual gases bound up in crude oil (ethylene, propylene, butadiene, among others) ascend and arrange themselves in discrete layers. Then they're drawn away into the dome-shaped storage structures, later to be polymerized or put to other chemical uses. I ponder plastic's genealogy, a family tree germinating billions of years ago, with roots twining through first life, its trunk rising through the ancient world and branching through hidden alchemy labs and nineteenth-century kitchens, scaling ivied towers, fruiting in research divisions, seeding store shelves and oceans, always bending to man's hunger to know, build, discover, decode, compete . . . to solve problems and to get rich. Plastic flotsam is the end product of an eons-long chain of transformations beginning with the planet's earliest life-forms in the oceans . . . planktonic creatures and algae living and dying over billions of years, blanketing the sea floor in numbers beyond our ken. The still-settling earth folds over on itself, trapping these vast deposits in pockets, pressure cooking them into viscous black gravy loaded with hydrocarbons. In a sense, our plasticized ocean represents recycling at its most epic, and worst.

Before the internal combustion engine, petroleum, literally "rock oil," was only marginally useful. In places where petroleum seeped, man found things to do with it. The walls of Jericho were mortared with pitch, ancient Baghdad's roadways asphalted with it, and Chumash Indians in coastal California used it to seal their canoes. Marco Polo's thirteenth-century route to China took him through Azerbaijan, south of Russia, where a "fountain" of oil "could fill many boats each day." (Today, it does.) He helpfully noted that oil was not only "good to burn" but an effective treatment for camel mange.

In the 1830s, Yale chemistry professor Benjamin Silliman was the first American to distill petroleum into clean-burning kerosene. Oil seeps had been a contaminating nuisance to settlers digging water wells. But whale oil prices were rising and new illuminants held commercial promise. An enterprising New York lawyer, George Bissell, got wind of a seep in Titusville, Pennsylvania. In 1859, he commissioned a study from Silliman's son, Benjamin Jr., also at Yale. Silliman Jr.'s fractional analysis discovered what oil was made of—mostly "organic" hydrocarbon compounds, with a few residual metals. Before his study, Silliman was a promoter of coal oil, another new distillate, and skeptical of rock oil. Afterward, he was bullish on petroleum, but for two uses only: illumination and lubrication of machinery.

Nearly seventy-five years after Benjamin Silliman Jr.'s report, Standard Oil research director Carl O. Johns told a group of chemical engineers: "Petroleum affords a veritable mine of organic compounds. . . . Who can predict how far reaching both from economic and scientific aspects would be the results of a thorough scientific investigation of this almost 'virgin field'?" At this time, cars had begun their conquest of American roadways—23 million by the late 1920s—but Johns knew how many petroleum by-products were going to waste.

Now let's go even further back in time and meet three men who had something in common: December 29 birth dates, making each a tenacious, methodical, ambitious Capricorn. They are Charles Macintosh, a Scot; Charles Goodyear, a Connecticut Yankee; and Alexander Parkes, a Brit.

Macintosh, born in Glasgow in 1766, inherited his father's bent for applied chemistry, namely fabric dyeing. The more entrepreneurial Charles strayed into other areas, and Glasgow's new gasworks, which produced coal gas for illumination, caught his attention. Coal-processing by-products included ammonia, which Macintosh used in dyes, but also a tarry residue that was burned off in flares. Scot that he was, Mackintosh wondered if the tar could be put to good use. Thus, he learned to distill oily, volatile naphtha from tar. This solvent, as he already knew, acted on a New World import that was causing great excitement in the Old World: rubber. Derived from several equatorial trees, rubber was first collected during a 1735 French expedition to Peru. Indigenous people had already harnessed its natural elastic properties, making footwear and balls for sport.

As is always the case with breakthroughs, Macintosh proved the right person at the right time. Around 1820—before the national rail system—the British walked, rode horseback, or traveled in carriages, exposed to the isles' damp and punishing elements. With his deep knowledge of textiles, Macintosh got the idea of softening rubber with naphtha and spreading it in a thin layer between sheets of fabric. The result would be rubberized, waterproof cloth. And it was, but the early product was imperfect. Leaks sprang along stitches, wool oil degraded it, extreme heat made it mushy, and harsh cold embrittled it. And it smelled bad. By 1830, Macintosh had teamed with Thomas Hancock, inventor of a "masticator" that chewed up rubber tailings. The result was a tamer, less smelly rubber that spread smoothly over fabric and became a waterproof brand: Mackintosh.

Charles Goodyear was born in New Haven in 1800, thirty-four years to the day after Macintosh. Goodyear's hardware store in Philadelphia prospered, then declined, along with his health. Debts landed him in prison, but not before he'd developed something of a rubber fetish. He later wrote of it: "There is probably no other inert substance which so excites the mind." He wasn't the only one thus afflicted. A sort of rubber bubble—"rubber fever"—gripped the 1830s. Tons of the material were imported from South America to New York and Boston. But Goodyear ran into the same problems Macintosh encountered: rubber was sticky when warm, brittle when cold, and nasty-smelling. Goodyear vowed he would find a way to solve these problems.

During his detention, Goodyear asked his saintly wife, Clarissa, to bring him rubber samples to handle and study. Using his wife's rolling pin, Goodyear worked in a talc-like powder, hoping to temper its stickiness. The results were promising. Back home, he, his wife, and their small daughters fashioned the stiffened rubber into rain boots. But summer came and the boots melted. Clarissa's brother visited, warning Goodyear that his family had suffered enough, and for naught. Rubber was "dead," he claimed. Goodyear reportedly replied, "I am the man to bring it back."

After more trial, error, and a few glimmers of success, Goodyear's first true breakthrough came through the accidental agency of nitric acid. Goodyear had run out of raw rubber and decided to recycle a boot he'd gilded with bronze. Nitric acid would dissolve the bronze, he thought, leaving clean, raw rubber. Instead, the acid charred and hardened the rubber. He made more boots. Again they wilted. But Goodyear had learned something from rubber's reaction to the oxidizing sizzle of nitric acid. He teamed with a Massachusetts man, Nathaniel Hayward, who'd compounded raw rubber with sulfur and observed changes. Legend has Goodyear working in his home kitchen and dropping a chunk of sulfurized rubber on the hot stove. The combination of sulfur and heat at last produced a material with all the right properties: moldable, yet stable when cured. It was 1839.

Nature abounds with natural polymers. Bone, horn, shell, hair, fingernails, wood, proteins, even DNA are polymers, meaning made of molecule chains. In natural rubber, the polymers shimmy. That's what makes rubber bouncy. Unknowingly, Goodyear invented a process that caused the polymers to "cross-link"—to bind to each other three-dimensionally. Meanwhile, back in England, Thomas Hancock, Charles Macintosh, and another associate, William Brockedon (credited with the term "vulcanization"), continued their attempts to coax rubber into a more stable form. They'd gotten their hands on a sample of Goodyear's solid rubber by 1842, but had they decoded it? Goodyear waited till January 1844 to patent his process in England. Hancock had filed in December 1843. Goodyear traveled to England when Hancock sought to prove he deserved the English patent. In court, Hancock's side called expert witnesses who testified that Hancock could not have deduced Goodyear's process by looking at his sample. Though Goodyear's American patent stood, his English application was denied and, with it, the wealth that always remained just beyond his grasp.

Curiously, support for Goodyear's claim was later lent by our third December 29er, Alexander Parkes, born in Birmingham in 1813. Like Goodyear, Parkes began his career working metals. His first patent was an electroplating process for fragile objects. So special was this process that Prince Albert visited Parkes's metalworks and was delighted to receive a silver-plated spiderweb. Two years after the Hancock-Goodyear patent case, Parkes invented a "cold cure" process for rubber that made vulcanization quicker and cheaper. At Hancock's urging, Macintosh & Co. bought the patent for five thousand British pounds and quickly saw vast cost benefits. During negotiations, Hancock told Parkes he *had* analyzed and learned from Goodyear's sample. Parkes made note of the admission for posterity.

Though the bulk of Parkes's sixty-plus patents were metal-related, he's accorded the title "father of plastics." Parkes began working collodian, a gel that dried as a clear film, made by dissolving natural fibers in nitric acid. Doctors had already adopted it to seal small wounds. His first thought, inspired by his dealings with Macintosh, was to develop a new type of waterproof material by impregnating fabric with collodian. But another enticing challenge lay before him. Billiards, at this time, was all the rage throughout Europe and America. While players of modest means cued up with wood or clay balls, the rich favored ivory, which, like horn, is a natural polymer. But ivory was costly, and the ivory trade had led to a lamentable scarcity—of elephants *and* ivory.

Aiming to create a man-made billiard ball, Parkes started from scratch, first dissolving cotton and ground wood fibers in nitric and sulfuric acids. Into this paste he blended aromatic vegetable oils derived from the castor bean and wood—natural naphtha. The result was a translucent dough that could be molded, textured, and colored to simulate ivory and horn. It dried hard and glossy. Parkes realized his new material, dubbed Parkesine, held potential far beyond the billiard table. At London's 1862 International Exhibition, he won a bronze medal for molded items such as knife handles, pipe stems, and medallions, some resembling metal, others natural materials like shell. Parkes had big plans for Parkesine: brushes, shoe soles, whips, walking sticks, buttons, brooches, buckles, decorative items, umbrellas, countertops, and, of course, gaming balls—items often made from plastic today. Cheaper than leather, rubber, or horn, Parkesine radiated commercial potential. But the Parkesine Company sputtered and failed, reportedly due to Parkes's cost-cutting use of cheap raw materials. That Parkes's billiard balls had a tendency to explode was another drawback. Thus, Parkesine proved but an evolutionary step. An associate took over the faltering business and had modest success producing cuffs and collars. But in an era when open flames were part of daily life, he, too, had a problem. His cuffs and collars didn't merely catch fire but burst into flame. For that matter, rubber processing and plastics factories proved to have the same combustible tendencies.

It was an American who took Parkes's inventions and transmuted them to lucre. This was John Wesley Hyatt, born in 1837 in upstate New York. Like most nineteenth-century inventors, he was not university trained but instead apprenticed in a trade, in his case printing. His first patent—of 228—was for a home knife sharpener. In the 1860s, the largest U.S. maker of pool tables and equipment, New York–based Phelan & Collender, offered the prodigious reward of $10,000 to the inventor of the elusive synthetic billiard ball. The game's popularity at the time was such that one contemporary report claimed professional billiards matches were followed more closely than Civil War battles. Historical accounts differ as to whether Hyatt licensed Parkes's collodian formula or had even awareness of Parkes or Parkesine. At the time Hyatt took up the challenge, Parkes was setting up his star-crossed factory. But Hyatt claimed ignorance of Parkes's work in 1914 when he collected applied chemistry's most coveted award, the Perkin Medal.

In the 1870s and into the next decade, bitter court battles were fought between Hyatt and Parkes's successor, Spill. While Parkes was credited with the invention of Parkesine, Spill was denied patent rights and soon failed. But Hyatt and his brother, Isaiah, were permitted use of the formula and were wildly successful with their version, dubbed Celluloid by Isaiah. As to the reward, the historical record is murky. Phelan & Collender may have judged Hyatt's new material too similar to that of Parkes to deserve a prize. In any case, the Hyatts built a factory in Newark that soon caught fire. A new factory in New York added to the product line with denture plates, piano keys, and personal-care items.

Macintosh, Goodyear, Hancock, Parkes, and Hyatt all started with materials found in nature—tree sap, cotton fiber, wood "flour," and bone dust—and basically cooked them with chemicals that changed them into something new. The results were "semi-synthetics" that have endured in products like rayon fabric, cellophane, Ping-Pong balls, rubber bands, and flip-flops. The proto-plastics were used to make goods that were kept, not used once or for a year and thrown away. Still, forward-looking entrepreneurs with a chemical bent clearly saw the commercial potential in cheap man-made substitutes for distant, hard-to-get natural materials. But synthetic resins made entirely from chemicals would have to wait till the new century. The term *plastic* was in use by century's turn, but the true chemical nature of polymers was yet to be decoded.

Leo Baekeland was a new breed of plastics pioneer. Belgium born, he was university trained, a summa cum laude Ph.D. intent on monetizing his chemistry expertise. He chose for this reason to honeymoon in the United States, wangling a job in a New York photographic house. In 1893 he invented a new type of photographic paper that produced images in artificial light, named it Velox, and sold it for $1 million to George Eastman of Eastman Kodak. With the proceeds, he could have retired to the estate he bought, overlooking the Hudson. Instead, he equipped it with a lab. Baekeland's idea was to invent a cheap, synthetic form of shellac, then used to insulate wiring. Natural shellac is the resinous secretion of the female lac beetle, deposited on the bark of an exotic tree in Asia. Demand exceeded supply, making it expensive.

Baekeland combined carbolic acid, a mild acid distilled from coal tar (now known as phenol and derived from petroleum and natural gas), with formaldehyde, a derivative of good old naphtha. Both toxic, both hydrocarbons. He blended them with a catalyst and pressure-cooked the slurry in a sealed, paddle-equipped capsule the size of a domed barbecue. His staff dubbed the contraption "Old Faithful" and knew to stand back when Baekeland added catalyzing alcohol. True to form, several fires erupted, prompting Old Faithful's relocation from lab to garage. After years of trial and error, Baekeland produced a viscous moldable resin that dried to a hard, glossy sheen, exactly what he had been looking for. Baekeland presented an array of attractive Bakelite objects—pipe stems, buttons, bangles—at a 1909 gathering of the American Chemical Society, nearly a decade after beginning his experiments, and caused a sensation.

Bakelite did replace lacquer as an electrical insulator in industry, and much more. The nascent auto industry quickly latched on to Bakelite for knobs, steering wheels, and door handles, as did makers of home appliances like toasters, electric irons, and vacuum cleaners. Bakelite pens and costume jewelry are now collector's items, as are vintage Art Deco radios. By the early 1920s, when Baekeland's factory was churning out 8.8 million pounds of plastic each year, people knew an "age of plastics" had arrived. Baekeland was even featured on *Time* magazine's cover. No shrinking violet when it came to promotion, Baekeland himself called Bakelite "the material of a thousand uses." Until the 1960s, rotary phones were made of Bakelite. We had two. The finest billiard balls are still made of phenolic resin, the generic term for the plastic Baekeland invented. If Bakelite collectibles were floating around in the gyre, they could pay for a good chunk of the voyage. No chance, however. Bakelite, a dense material, would sink to the ocean floor.

Bakelite and celluloid are of a category of plastic called "thermoset." They harden when cured and, when heated, scorch instead of melting. Thermoset plastics are rigid, not flexible, because the polymer chains they're composed of cross-link three-dimensionally. These days they're mostly used to make durable goods like computers, dashboards, helmets, glasses, strollers, and surfboards. They are distinct from generally more flexible "thermoplastics," which melt when heated and can be formed and re-formed. In 1939, a few years before Baekeland's death, chemical giant Union Carbide bought Bakelite. By this time Baekeland had surrendered to extreme eccentricity, eating alone from tin cans and fighting with decadent offspring. He was spared a lurid, if tragic, scandal in the 1970s involving . . . well, rent the film *Savage Grace* if you must know.

Somewhat surprisingly, factories were whipping up synthetic resin products by the ton before anyone knew exactly what a polymer was. The early polymer pioneers were flying blind, achieving predictable results based upon trial, error, serendipity, and chemistry basics—like instinctive cooks. But once polymers were explained, by a German chemist in 1920, the new model revolutionized thinking about plastics and spawned a new science: polymer chemistry. If you had a grasp of the concepts and a background in organic chemistry, you were ready to try your hand at making a synthetic polymer.

Let's have a closer look at how petroleum processing works. It operates on two tracks. One produces petroleum products, the other petrochemicals. Of a forty-two-gallon barrel of oil, 42 percent will be used to make petroleum products, basically gasoline, burnable fuels, lubricants, and asphalts. These are derived from the heavier fractions of crude oil. The rest, lighter fractions, are "cracked" into individual hydrocarbon compounds. Of these, there are two basic categories: the olefins, meaning "oily," and the aromatics, which are more volatile compounds. Thermoplastic feedstocks, also called nurdles, or preproduction pellets, are produced from both categories.

Harvesting these hydrocarbons from crude oil is something like reverse engineering. Imagine you baked a cake but it turned out badly. Even the dog's not interested. A disappointing, low-value result. But let's say there was a way you could unbake the cake and reclaim the separate ingredients to make other things. Such a process would be akin to petroleum cracking. You'd put it in a cylinder, add heat, a chemical catalyst, and/or pressure, and whirl it all around. The heavier ingredients would migrate downward as the lighter ones waft upward. The chemical fractions are now vertically layered within the tower. Tubes attached up the length of the tower inhale the discrete constituents, sending them on their separate ways for further refinement and processing. Oil refineries process oil. Chemical plants make things like plastics and pesticides from petroleum by-products.

That's why where there's petroleum refining you'll likely find chemical plants within shouting distance, often linked by pipelines. Of the twenty largest petroleum refineries in the world, only six are in the United States, the largest being number six, in Baytown, Texas. India has the largest, Venezuela the second biggest, and South Korea the next two. But if you combine the capacities of the six refineries between Los Angeles International Airport and Long Beach, you have refining capacity comparable to the giants. The most prolific petroleum by-product is ethylene. Most of it will become plastic. But plastics are only part of the story. Other olefins and aromatics will undergo astonishing chemical, thermal, and catalytic manipulations to become myriad nonplastic products, such as paints, adhesives, rubbing alcohol, paraffin candles, cleaning fluids, furniture wax, shoe polish, pesticides, fertilizer, food flavorings, fragrances, industrial coatings, antihistamines, pill coatings, crayons, hair dye, fabric dye, ink, antifreeze, anesthetics, topical ointments, cosmetics, and shaving cream, to mention but a few.

I once wangled a tour of the ARCO—now BP—petroleum refinery at Wilmington, the Port of Los Angeles. What I saw was this: a lot of tanks, pipes, and ducts. What I learned was that they were making a crystallized polymer called fluff that resembles fake snow. Once residual catalysts and other impurities are removed, fluff is fed through a mill and becomes, of course, nurdles. The nurdles are shipped to all those tens of thousands of processors located pretty much everywhere, as flour would be shipped to bakeries.

Polyethylene, an olefin, in all its many guises is, by far, the dominant plastic type. The industry lobbying arm, the American Chemistry Council, reports current worldwide production of 90 billion pounds a year. Ironically, the first polyethylene was a bioplastic, discovered in 1933 by two English chemists named Reginald Gibson and Eric Fawcett at Imperial Chemical Industries. First they derived ethylene in several laborious steps from molasses. To this they applied 28,000 pounds per square inch of pressure and intense heat, then turned off the cooker and went home for the weekend. The following Monday, they opened the chamber to find a mere gram of waxy solid residue, which they recognized as a polymer. The scientists had trouble replicating the process until they thought to add in oxygen. This catalyzed and stabilized what had been a completely unpredictable reaction.

Imperial opted not to move forward with the new material, but Gibson displayed the polyethylene sample at a Cambridge gathering in 1935 and others grasped its potential. During World War II, the English used polyethylene (PE) to enhance radar performance and insulate cables. But for nearly two decades, PE remained a specialty industrial material. The material that would one day smother the planet with shopping bags and gallon milk jugs could not, in its early days, be cheaply or easily produced.

In the first half of the twentieth century, plastics development was stoked by corporate competition, research funding, and war. Much of European and U.S. chemistry in the period between the wars focused on creating exotic new synthetics to replace expensive natural materials sourced from distant equatorial regions. Between 1933 and 1939, chemical company labs in the United States and Europe—especially Germany—invented clear-film Saran (made of polyvinylchloride), acrylic, polyurethane, Lucite, and polystyrene, and they stumbled on Teflon.

DuPont is the oldest of today's multinational chemical companies. A century after its 1802 founding as a gunpowder manufacturer, a younger generation of du Ponts wrested control of the company and broadened its scope to include chemicals. It was well known by then that nitrocellulose, DuPont's primary product, could be catalyzed into celluloid. In the teens DuPont acquired companies that made celluloid and a Bakelite imitation. But it wasn't until the late 1920s that the company decided to set up an experimental lab at its Wilmington, Delaware, headquarters and staff it with the best minds they could lure from academia. One recruit was Wallace Hume Carothers, a young Iowa-born Harvard instructor, whose first breakthrough was neoprene, a fully synthetic rubber, still used today to make wet suits for surfers and divers. Then came nylon, a new kind of plastic, not cooked but condensed from a chemical reaction. The silky fiber, "as strong as steel, as fine as a spider's web," patented in 1937, had, by the following year, been made part of a personal-care product: the nylon-bristle toothbrush. Boars suddenly found themselves in low demand. Commercialization proceeded apace with wildly successful nylon stockings, causing riots in 1941 due to wartime shortages. Nylon became vital to the war effort, substituting for silk in parachutes, flight suits, and rope.

War demanded large-scale production of materials that could replace heavy, shattering glass, scarce metal, inaccessible rubber, and tropical plant fibers used for ropes. War offered the perfect proving ground for new materials and brought enormous government contracts to chemical and manufacturing companies. Even the first commercially produced ballpoint pen—becoming the Bic Cristal in 1950—was licensed by the RAF because it performed better at high altitudes than leaky fountain pens.

Each of the major chemical companies came to be identified

with a certain plastic or plastics. If DuPont's are nylon and neo-prene, Dow's is polystyrene (PS), another plastic with bio begin-nings. Styrene is a naturally occurring chemical, found most abundantly in the tropical styrax tree. A German apothecary in the 1830s had heated styrene and produced a jelly-like residue. Other chemists subsequently noodled with the material, and finally, at Dow, they combined styrene with the hydrocarbon butadiene. The result was the first fully synthetic rubber, an ob-vious war effort asset. Another form, crystal PS, is hard and brittle. We see it today as Bic disposable pens, lighters and ra-zors, fast-food utensils, CD and DVD "jewel boxes," and refrig-erator liners and bins. This isn't a plastic type of great ubiquity in the offshore marine environment because its density causes it to sink—with the notable exception of those hollow disposable lighters. But we do see plenty of the foamed or expanded form of PS, which is almost weightless. This material came about in the 1930s when a Dow chemist combined molten polystyrene with a gas to make a flexible insulating material. True Styrofoam was patented in 1944 and now has three basic applications: home insulation, crafts, and flotation and bumper devices for boating. Styrofoam is not to be confused with run-of-the-mill expanded or foamed polystyrene, which debuted in the 1950s in the form of hot-drink cups, take-out food clamshells, and packing pea-nuts. Because of its light weight, foamed PS shares a propensity with the plastic bag to be swept away. I see and capture it often, not only in small chunks—because it breaks apart more readily in sea water than other plastic types—but also in the form of errant buoys and fishing floats, typically pocked with piscine munch marks. The American Chemistry Council brags on its website that expanded polystyrene cups, plates, and food clam-shells have smaller carbon footprints than their paper equivalents. That may be, but its footprint is also more persistent and chemical laden than those of paper products.

By war's end, plastics had taken root, but they had yet to go mainstream. And while the demands of war had force-fed the chemical and rubber sectors of the economy, war's end threatened to pop the bubble. New markets and new uses for these miraculous man-made materials had to be found, and fast.

Pittsburgh Plate Glass Company was stuck with a tank car containing 38,000 pounds of an acrylic-like plastic, still in a liquid state, patented under the name CR-39. The lightweight but strong material had been used to make fuel tanks for bombers, extending their range. The company reached out to other industries, hoping for brave new customers willing to give this novel material a whirl. The company finally hit pay dirt with the optical lens industry. The problem of heavy glass lenses was soon solved. Even today, a substantial percentage of glasses, especially dark glasses, is made from CR-39.

The turning point for polyethylene came in the early 1950s, soon after Phillips Petroleum set up a research division at its Bartlesville, Oklahoma, headquarters. Two chemists, J. Paul Hogan and Robert Banks, were charged with finding out if oil refining by-products—ethylene and propylene—could be turned into performance-enhancing gasoline additives. Their experiments included adding a metallic catalyst to propylene. The unexpected result was a crystal polymer. They similarly catalyzed ethylene and got the same result. It was the metal catalyst that literally changed the world, transforming PE from a specialty item to one of the most ubiquitous materials ever produced. Prewar chemists could not have imagined that their efforts would yield an unstoppable avalanche of nonessentials.

Phillips's began cranking out PE without a clear marketing

plan, resulting in warehouses filled with the stuff. The product that saved Phillips's plastics division and launched PE into American culture was Wham-O's Hula Hoop, the hollow ring based on a bent bamboo toy from Australia. The Hula Hoop was launched in July 1958. By year's end, four million units had sold. Two years and 25 million Hula Hoops later, Wham-O was sitting on profits of $45 million. The high-density polyethylene Frisbee had actually preceded the Hula Hoop by a year. The pie-tin-shaped toss toy was no match for the hoop in faddishness, but it had legs. As of 2010, Wham-O, now Chinese owned, estimated that 200 million Frisbees had been sold. So it was toys that ushered in the Plastic Age. And for a while, plastics were mostly used to make consumer goods and industrial parts, not packaging or disposable products.

And so it was that at first, plastics seemed exciting and wonderful. They were so Tomorrowland, the signature material for adventurous Space Age citizens. I well recall visiting Disneyland soon after its grand opening in 1955. There, at the far end of Main Street, not far from Sleeping Beauty's Castle, was Monsanto's all-plastic house of the future that seemed very fabulous, and easy to clean! The house is gone now; the oxidizing effects of Anaheim's sun and smog would have degraded the polymers, leading to irreparable cracks. Plastic polymers may last nearly forever, but plastic things don't hold up very well, often not as well as the natural materials they simulate.

Then, at some point, plastics stopped seeming cool and new and began seeming false and tacky. It had already happened by the late 1960s when audiences sneered at the family friend who exhorted young Benjamin Braddock in *The Graduate* to seek a career in plastics. Yet this was still well before most disposable plastics were in use: before PET beverage bottles, frozen microwave meals on disposable plastic plates, and flimsy plastic shopping bags. But we were already resigned. The genie was out of the bottle. We stopped noticing plastic, and it seemed to disappear even as it proliferated.

Since 1976, the tipping point, plastic has been "the most used material in the world," according to the American Chemistry Council. But now it is mostly used for packaging, with building materials—insulation, PVC siding, and synthetic carpet—coming in second, before consumer goods like toys and computers. Plastics used to be a strong export item for the U.S. economy, generating a $10 billion trade surplus in the 1970s. Now, like a Ziploc dropped on a hot burner, the U.S. plastics industry has shriveled, with Asia taking up much of the slack. Now we import more plastics than we make. Would that global plastics production were declining, but this is decidedly not the case. Current estimates put worldwide plastics production at 300 million tons a year, fifteen million tons more than the annual world consumption of meat, an almost inconceivable amount, especially when you consider that meat is eaten and digested, but plastic decays very, very slowly and is constantly accumulating.

Where will it all end? My oceanographer friend Curtis Ebbesmeyer wonders if we can "shut off the plastic switch." In the day's waning light, my visitor and I take a last look at the vast complex of complexes where oil becomes chemicals and chemicals become polymers and head back to the world of "end users." Shutting off the switch to this juggernaut will be a herculean task. Turning down the volume seems a good place to start.

CHAPTER 3

SURFING THE LEARNING CURVE

THE PERSON WHO RETURNED from that very long voyage in 1997, the one who noticed plastic scraps in the middle of the North Pacific, didn't know much beyond the basics about plastics' ocean pollution. But he—that is, I—was about to learn.

We're sea-wobbly but cheerful when we finally step on dry land after that protracted crossing. We go our separate ways, meaning the crew is swept away by family and I cross the street to the house I grew up in and now call my own. I can keep an eye on *Alguita*, now sitting quietly in her slip, from nearly every room. I resume my terrestrial routine. After two months away, there's much catching up to do. I set to work in my yard, my subtropical Eden, an arboreal retreat shaded by exotic *sapote negro*, cherimoya, guava, banana, and papaya trees, and planted with organic vegetables and herbs that feed us well. All this on a compact marina lot. I stop by the thriving community gardens we established in empty lots in downtown Long Beach, in the aftermath of riots and arson in the early nineties. I catch up on paperwork, tell my stories about the voyage, and write a report for the *Algalita Newsletter*. I warn in this report that floating plastic trash may be on its way to becoming "our oceans' most obvious surface feature."

Alguita's had her share of South Seas adventure and proved her mettle in a demanding transoceanic race. Now she's well shaken out and ready to settle into her role as a full-fledged research vessel. We'd removed five hundred meters of wire cable from her trawl winch to lighten her for racing. Now it must be reinstalled. Stressed from the race, her rigging needs repairs, as do the sails, which we bundle up for delivery to the local sail loft. And there was the matter of *Alguita*'s insurability. After our South Seas misadventures, Lloyd's of London declined to renew her policy. To facilitate getting her reinsured, I decide to get myself licensed by the U.S. Coast Guard as sea captain. The requirements are stiff, but I've already spent more than the necessary 365 "days" (minimum four hours each) at sea, including time aboard the *Sea Watch*, a research vessel operated by the Southern California Marine Institute, which allows me to go for my 100 ton license. Along the way I've absorbed knowledge about tides, winds, charts and boat mechanics, in no small part thanks to the United States Power Squadrons' excellent boating education courses. But I opt to take the USCG training class offered by Orange Coast Community College before sitting for the punishing written test. I eventually pass (had to take the "Rules of the Road" section twice) and then the medical and drug tests. I'm now Capt. Charles Moore, U.S. Merchant Marine Officer and Master of Steam, Motor or Sail Vessels of not more than 100 Gross Tons, personally responsible for the safety and welfare of my crews.

Still tugging at my mind, however, is this image of plastic trash

and scraps fluttering like moths in liquid air for miles on end in
the still center of the Northeast Pacific.

I'M SOMETIMES ASKED if a sudden epiphany set me on course
back to the northeast subtropical Pacific, where I'd seen so much
plastic. I can't say that was the case. I do know that business as
usual does not dim thoughts about the plastic soup I'd seen out
there. The notion of sailing back for another look probably began
to percolate during the 1997 voyage, when I doodled my estimate
based on half a pound of plastic per one hundred square meters.
This gave me a ballpark figure in excess of six million tons of
plastic debris in the Northeast Pacific between Hawaii and the
West Coast, an area of about two million square miles. It's a
situation that seems well worth further investigation. Curiously,
events soon to follow seem to channel me back to the deeps and
a closer look.

I CONSIDER MYSELF a man of science, but at times serendipity
seems to guide my way. I read an article in our local paper about
a meeting being called by one Dr. Stephen Weisberg. He's described as the dynamic head of a state agency charged with quantifying pollution in Southern California's coastal waters. A man
after my own heart. I'm inspired to write a letter to the editor,
noting how lucky we are to have Weisberg working to improve
the health of our coastal waters. And I mark the meeting on my
calendar. It's the fall of '97, a couple of months after that first voyage through the gyre.

    Weisberg's agency is the Southern California Coastal Water
Research Project, affectionately known as SCCWRP, pronounced

"Squerp." The agency is publicly unveiling a research plan for a study called Bight '98—a bight being an area of circulating sea water between prominent headlands. The purpose is to check contamination levels in the coastal water system between Point Conception and Punta Banda and to compare it with measurements from an earlier bight study.

The Southern California Bight runs about one hundred miles north and south of Los Angeles—from northerly Santa Barbara south to upper Baja California. The study is an ambitious project that calls for seasonal samples to be taken at 416 sites. This isn't science for science's sake. SCCWRP aims to find out if stiffer pollution regulations have resulted in better water quality and a healthier coastal ecosystem.

Public health and safety are not small concerns for a coastline that draws 175 million visitors annually. If beach waters are polluted, local businesses suffer, as do city, county, and state tax revenue coffers. Given these shared concerns, SCCWRP was formed by a consortium of public health and water quality agencies in the late 1960s. At this time there was no denying that effluents generated by Southern California's surging population were turning coastal waters into a witch's brew of biota-killing chemicals. The Clean Water Act and the EPA were yet to come in 1972, but California is always ahead of the game.

Again, Rachel Carson's influence cannot be ignored. Remember, it was already 1962 when *Silent Spring* warned of possible extinctions resulting from injudicious use of the pesticide DDT. In 1947 the Montrose Chemical Corporation of California, located near Los Angeles in the coastal town of Torrance, had begun producing DDT. It was soon to become the country's top supplier. As a child—and before in my mother's womb—I was exposed to DDT mosquito spray in the canvas tent the family used for frequent camping trips to Baja California. Mosquitoes didn't have a chance, but I can't help but wonder if my few physical anomalies resulted from my mother's exposure. In 1972, DDT was banned in the United States, but Montrose continued to make and export the chemical for another decade. A veritable Vesuvius of toxics, Montrose also churned out PCBs, another class of insidiously persistent synthetic molecule, used widely for decades to lubricate, insulate, and flameproof industrial equipment and building materials before being banned in the late 1970s. Montrose estimated it had discharged 1,700 tons of DDT into the county wastewater system between the late fifties and early seventies. The effluent emerged at White Point, atop which sits the upscale community of Palos Verdes Estates. Montrose also copped to discharging at least ten tons of PCBs. A broad swath of the coastal shelf was and remains a marine "sick zone" and would eventually become an EPA Superfund site.

Montrose wasn't the sole reason for SCCWRP's formation, just the most glaring. The new agency's mission was to scientifically quantify and qualify threats posed by biological, mineral, and chemical contaminants pouring from the L.A. Basin watershed into near shore waters. The point was to let sound science guide policy pertaining to water treatment and industrial discharges. The joint powers charter in 1969 provided for a three-year study, but SCCWRP performed so ably that its funding agencies have kept it going to this day. SCCWRP remains vigilant, making the Southern California Bight the most studied marine zone in the nation and, no doubt, the world.

I showed up at the Bight '98 meeting wearing my white Power Squadron dress shirt with navy epaulets. As you probably don't know, this shirt, with navy slacks, is the official garb of a nearly one-hundred-year-old national nonprofit organization called the

U.S. Power Squadrons. In partnership with the U.S. Coast Guard Auxiliary, the USPS provides education and training in boating safety, and I gladly fulfill my training duties when called upon. During the meeting, I contributed my several cents to the discussion and was approached afterward by the boyish Steve Weisberg. I found myself willingly conscripted into the Bight '98 water-sampling effort, along with ORV *Alguita*. We'd be one of twenty-one organizations cooperating in the sample-gathering effort. My, and *Alguita*'s, special niche would be to serve the Mexican scientists who lack access to a research platform for taking samples south of the border. I'd also translate for Spanish-speaking partners from the Autonomous University of Baja California.

The study went well and yielded encouraging results, but the major outcome for me was entrée to the world of Weisberg—a master at connecting science-based environmental monitoring with policy—and his clever colleagues. I like doing research, and I see that it can matter. So I think harder about sailing back to the mid-Pacific, where I saw so much plastic garbage. This is ten years out from MARPOL Annex V. Why is there so much plastic trash out there? And is there any reason I can't be the one to quantify it?

Before I resolve one way or other, I stop in at the Costa Mesa offices of SCCWRP. It's no small effort to mount a three-week voyage to a mid-ocean dump site, a thousand miles west of the West Coast. If I do go, the science has got to be unimpeachable. The marine monitoring I've done thus far has been as a contractor, and it's all been coastal. I wonder if Weisberg and his staff will be able to figure out how to design a study plan for the mid-ocean, which is as different from coastal waters as Kansas is to Hawaii. I'm also aware that SCCWRP has researched marine debris and beach trash, not just chemical and biological water quality. Weisberg tells me about a study they ran in the early 1990s. It had been designed to quantify cigarette butts in beach sand, but wound up finding far fewer butts than plastic bits, many of them tiny perfect pellets known as "nurdles." They were floored. SCCWRP also studied debris in the coastal seabed and found plastic trash there as well. It feels like a good match.

To this meeting I bring the daily weather faxes—alerts issued by NOAA's Point Reyes weather station—we received during the 1997 Transpac return. I'd been poring over them myself, searching for clues as to why this calm, massive, avoided part of the North Pacific would be home to so much plastic trash. What impressed me about these charts was the unvarying presence of a stable area, a high-pressure zone that never dissipated though it tended to slide around. Wind and ocean currents swirled around this zone, but the High remained calm and stable, something like a Buddha in Grand Central Station.

Once settled in a conference room with Weisberg and his statistician, Molly Leecaster, I lay all these weather faxes out on the floor. I order them sequentially so they can see this phenomenon of the North Pacific High, how it hovers and glides over this one area, about halfway between Hawaii and the West Coast. I say that it seems to me there may be a link between this stable marine high-pressure zone and the strange floating dump site. Little do I know that I've "discovered" something oceanographers had been investigating for more than a decade. They call it a gyre, a term I was soon to hear for the first time. No matter. It was what it was: an atmospheric phenomenon with characteristics so distinctive that it almost seemed like a fixed climatic feature.

The question I pose to Weisberg and Leecaster is basic: How do you take rigorous scientific samples from an area so incredibly vast—yes, twice the size of Texas—and devoid of physical boundaries? The task was very different from the sampling we'd done in the Southern California Bight, where coastal waters are naturally defined by contiguous land features. In the open ocean, you're dealing with a different beast. Out there, a single body of water becomes many, each loosely bounded by interlocking atmospheric pressure systems and ocean currents, none absolutely fixed. Coastal waters are powerfully influenced by land features. Pelagic waters — the open ocean — yield to the forces of currents. Mid-ocean currents take cues from mostly inconstant variables like air pressure, wind, temperature, salinity, lunar position, and the Coriolis effect that results from the earth's rotation. Because air pressure systems move around, the high would be sliding around as well, dragged north, south, east, and west like a cursor on a computer screen. Ocean current charts and other physical information about the mid Pacific would be of little use. We'd have to target the dump site, the study site, using atmospheric information. So the challenge is to extract scientifically rigorous random samples from a moving target. I say to Weisberg and Leecaster that it seems this high-pressure zone could be considered a stratum, which is the physical marker used in shore water studies. Maybe we can collect samples from a random line within that stratum. So this is how we develop a random trawl pattern for a study of plastic content in the North Pacific Subtropical Gyre. Because by now I realize I have decided to go for it.

Leecaster works up a map showing random lengths for each trawl and random distances between trawls. The positions we chart lie within an area about the size of Wisconsin. One transect is east to west and the other one north to south, so we'll get defensible and representative samples for an area that's in constant motion. The plan calls for us to sail to the central cell of the high-pressure zone, the doldrums, and start our sampling there.

I learn of one precedent for a study such as mine. It's in SCCWRP's library, and it's from 1984, conducted by Alaska-based researchers R. H. Day, D. G. Shaw, and S. E. Ignell, who were trying to figure out why so much derelict fishing gear and other plastic trash was washing up on parts of the Alaska shoreline. They'd trawled debris from a Japanese fishing boat in the Northwest Pacific between Asia and Hawaii and set up buoys that collected floating trash. They hadn't ventured much into the eastern Pacific, where I intended to go. Despite publication in a science journal, the findings hadn't made the news, which surprised me. It's as if the science exists in a transparent bubble, its message muted.

I present the Algalita board with a proposal. The board unanimously agrees that a mid-ocean research expedition supports the foundation's mission and could well produce research results of considerable importance.

I have approval. I have a research plan. Now I need a crew. I think of Robb Hamilton, an ornithologist and author of *Rare Birds of California*, who's managed to parlay his passion for birds into a successful consulting business. Our eco-paths have crossed here and there, and I'm impressed by his depth of knowledge. Maybe he'll appreciate the chance to observe seabirds rather than the usual California gnatcatchers and cactus wrens. I'm also aware that seabirds are plastics eaters, but more to the point, I like working with birders. They have exceptional powers of observation.

Mike Baker, my neighbor across the alley, is a retired highway patrolman who also happens to be an eco-activist and Congressional Cup sailor. He'll be first mate and prove a tremendous asset on the crew. He suggests John Barth, a former head lifeguard with a distinctive claim to fame. He was there at the moment when junior lifeguards at Huntington Beach coined the term "nurdle" for those little plastic BBs in the sand. (These were among the finds in SCCWRP's beach trash study.) It was the 1970s. The junior guards were sitting on the beach, waiting turns to test their lifesaving skills, idly sifting plastic pellets from the sand, when one of them referred to the bits as "nurdles." It was too perfect not to catch on. Barth will later tell me how they theorized about the nurdles' origins and uses. Their favorite was ingenious if not entirely plausible. It had crews sprinkling nurdles on cargo ship decks so heavy objects could be scooted around more easily. (Never mind that the crews would also be slip-sliding around.) Afterward, the theory went, the crew would sweep the nurdles into the harbor, where they'd catch currents and tides to the shore. It will take a while to learn the truth about nurdles, that they are "preproduction pellets," base synthetic resins that emerge by the trillions from central chemical plants, then get shuttled to tens of thousands of processors in all the populated corners of the world. They are industrial products, never meant to be seen by a beachgoer or found bobbing on deep ocean waters. But they easily escape the distribution system and go feral, with billions eventually winding up in waterways and the oceans.

IN JANUARY OF 1999, serendipity strikes again. I get a call from a stranger in Hawaii. His name is James Marcus and he lives in Waimanalo, an idyllic beach town on the far side of Diamond Head. The Honolulu Coast Guard has given him my telephone number because of work I'd done with the Long Beach Coast Guard's Marine Safety Office in the early 1990s, before founding Algalita. Marcus and I wind up speaking on the phone for about half an hour. He's clearly a very spiritual guy. He tells me he meditates daily on the beautiful beach and considers Hawaii a sacred place. Some months earlier he'd realized he was half-lotusing on a bed of plastic bits. He begins to take note of the granules. They seem to be cyclical. Some days the deposits are negligible, on others, bountiful. He's upset, thinking deliberate dumping is occurring on or just offshore from this clean, perfect beach. He'd sifted a sample into a gallon Ziploc and taken it to the Honolulu Coast Guard station. The Coast Guard is charged by law to investigate reports of marine pollution occurring in the sovereign waters that extend two hundred miles from any U.S. coastline. But some cases rate higher than others, and clearly the Coast Guard considers Marcus's find to be innocuous. Hence the Coast Guard referral to me. I ask Marcus to send me his bag of plastic bits and he does. When I remove the Ziploc from the mailer, I see multicolored plastic chips that look to me as if they'd been ground up as part of the recycling process. But I am also reminded of what I saw on that last voyage, though many of these pieces wouldn't have been visible from my perch on deck. So, my working theory is that Marcus's bits are spill from an oceangoing container filled with flaked recycled plastics on their way to be reprocessed. I am soon proven wrong.

YET MORE SERENDIPITY. In April 1999, the *Los Angeles Times* runs a front-page feature about Curtis Ebbesmeyer, a Seattle-based oceanographer. He runs an informal network of beachcombers who help him chart ocean currents by reporting flotsam finds. Bill Wilson, then president of the Algalita board, sees the article and brings it to my attention. He remarks that Ebbesmeyer may be the only Ph.D. dealing with plastic marine debris and might make a good research partner. I ask the foundation's education director at the time, Susan Zoske, to make contact with Ebbesmeyer.

Zoske, who possesses a formidable combination of efficiency and ardor, has him on the line before I can turn around.

When I speak with Ebbesmeyer, I'm entertained as well as enlightened. In 1996 he founded a nonprofit organization called Beachcombers' and Oceanographers' International Association and he puts out an irregular newsletter, *Beachcombers' Alert*. His far-flung cadre of beachcombers notifies him when they come across items like Nike sneakers and floating bath toys, including the famous yellow "rubber" (actually PVC) duckies from a 1992 container spill. The ducks and their fellow frogs, turtles, and beavers, marketed under the Friendly Floatee brand, were beginning to be legendary for the far-flung places they were washing ashore. Ebbesmeyer and his computer savvy ally, James Ingraham, were using location data about shoe and bath toy recoveries to refine their magnum opus: a computerized ocean current model called Ocean Surface Current Simulator (OSCURS).

Thus begins a lively correspondence that changes everything. I tell Ebbesmeyer what I saw on my 1997 voyage and estimate as much as 3.5 million tons of floating plastic in the high-pressure zone between Hawaii and the West Coast. It turns out Ebbesmeyer is a few steps ahead of me, having predicted a debris accumulation zone in this area based on the OSCURS model. Nonetheless, it's reassuring that my solitary investigations and musings have put us on the same track. He even has a name for this part of the ocean I plan to study. He calls it the Great Pacific Garbage Patch. Too bad he didn't copyright the phrase. To him, at this time, it's a theoretical place. He'd like to see proof that drifting flotsam actually gathers there.

Ebbesmeyer is also highly intrigued by James Marcus's find on Waimanalo Beach and asks to see a sample. I pack up and send him half of my stash, offering my theory that the particles are

flaked recycled plastics spilled on their way from Hawaii to a
mainland reprocessor. After inspecting the sample, Ebbesmeyer
agrees on one count and begs to differ on another. He writes in a
letter dated July 29, 1999: "I was struck by your estimates of the
trash circulating round the Pacific High. Then I studied the bag
of plastic bits from James Marcus (2.1 ounces with 982 pieces). I
think yours [the trash I saw in the North Pacific High] and James'
are linked in the . . . conceptual model I've had for some time."

From my observations based on the weather faxes, I'm already
familiar with the concept of the fixed yet mobile high-pressure
zone. Now Ebbesmeyer is telling me that my Wisconsin-sized
study area within the high is the eastern eye of a larger gyre formed
by a massive circulating current twice the size of the continental
United States. Moreover, this eye has a twin, another theoretical
garbage patch, situated on the far side of Hawaii, southwest of
Japan. And that's not all. Both eyes swirl within the massive circular current that encompasses the mid-latitude North Pacific.
It's made up of ocean rivers that sweep south along the West
Coast, west above the equator, north past Japan and Korea, and
eastward, grazing the Gulf of Alaska on its way back to the West
Coast. The oceanographer is leading me toward a eureka moment.

As Ebbesmeyer explains it, ocean debris tends to get sucked
toward the centers of the twin vortices. They're something like
giant toilet bowls, though there's debate as to whether the centers
of these sub-gyres are slightly elevated or depressed. There's also
a sizable area lying between the east and west gyres, called the
Convergence Zone. Fittingly, it's where currents converge, along
with their flotsam loads, which mingle and mass. During El Niño
years—which 1997 was—the convergence zone moves southward,
toward the equator, and actually engulfs Hawaii.

Here's the revelation: Ebbesmeyer thinks Marcus's plastic

grains are "spit out" from the Convergence Zone, sifted out cyclically by Hawaii's beaches, to the horror of people like James Marcus.

Ebbesmeyer contends that ultraviolet light and ocean chemistry granulate plastic debris, a theory I question. I think these plastic flecks are likelier to have started out as litter inland, been weakened by exposure to heat and light, then swept out to sea, where they're pounded to bits by wave action, especially as they break against the shore. We kick this can down the road. But I can buy his theory that what Marcus found in the sands of Waimanalo came from the same plastic-polluted waters I sailed through two years earlier. And I'm intrigued by Ebbesmeyer's claim that the visible plastic scraps and chunks I saw are "the tip of the iceberg." It's his belief that "micro" bits far outnumber and possibly even outweigh what the naked eye can see. He makes a few suggestions for tweaking the sample plan, but most significantly he recommends using a much finer mesh net than I'd planned for capturing samples.

So I'm getting *how* Marcus's plastic confetti wound up where it did. But I'm still not getting where it's from—the sources—or if it really matters beyond being a disturbing eyesore and a classic downside of "progress." To my mind, entangling plastic debris, the nets, straps, and lines, are obvious threats to wildlife. But this small stuff—I want to know more about the implications. What I saw on my Transpac return in 1997 now appeared to be but one brushstroke in an emerging portrait of throwaway plastics and their ubiquity on land and sea. Will diving deeper into this mystery yield answers, or at least clues, about where the plastic is coming from, and its impact?

Ebbesmeyer enthusiastically supports my mid-ocean research plan. He's especially keen for me to look out for spillage resulting from "a meteorological bomb" that struck in October of 1998—a massive mid-Pacific storm that battered three container ships and sent 411 cargo-laden containers overboard. The contents are unknown because international law doesn't require shippers to disclose container losses. Ebbesmeyer predicts that floating debris from the spill will have arrived in the gyre by late August, when I'll be there.

We're still a crew of four, which is manageable, but five is ideal. I invite Ebbesmeyer to be the fifth, but he demurs, saying his efforts are better spent monitoring the shoreline. He nominates Steve McLeod, an Oregon coast resident and artist who's rendered fantastical book covers for the novelist Ursula K. Le Guin. He's a key member of the beachcomber network. McLeod is legendary among flotsam fanciers for finding and matching dozens of pairs of Nike sneakers that washed up where he combs the beach, years after a 1990 container spill. He will prove an able deckhand and research natural.

The crew is now complete. But I still need to find a trawl net fine enough to trap the microplastics predicted by Ebbesmeyer to be more plentiful in the gyre than anything else. Ebbesmeyer has convinced me the half-inch mesh net I've secured to run across the hulls is too large to catch the stuff that will tell a very different story about plastic pollution. I'd anticipated catching things like toothbrushes, bottle caps, and plastic buoys. The scraps I'd seen in 1997 I'd estimated to be a half inch and larger—big enough to spot from *Alguita*'s deck. That there might be an unseen blanket of granular plastics covering the ocean surface is a game changer. I'm reminded of how it felt when, in third grade, I took my first look through a microscope at a seemingly clear drop of pond water that turned out to be teeming with tiny flagellating creatures invisible to the naked eye. Amazed, engaged, and a little unnerved.

CHAPTER 4

SWEPT AWAY: THE OCEANS AS GLOBAL DUMPSTER

HERE ON THE DECK of *Alguita* drips a broken blue plastic crate holding items that might have been culled in a children's neighborhood treasure hunt. A toothbrush, a toy car, a rubber sandal, a comb, bottle caps, a Popsicle stick, a shopping bag, all plastic. But they are faded and worn from years, maybe decades, in salt water. Most have been nibbled and look like chewed dog toys. They'd seem benign if I hadn't found them where they least belong, in the middle of the Pacific Ocean, about six hundred miles north of Honolulu. I've spent two hours rowing along a mid-Pacific windrow—more like a marine midden—plucking them from the water. Windrows, natural phenomena produced by wind action, used to look like bands of foam on the ocean surface. Now they are often bands of unnatural trash. The trove includes other items, many from the fishing fleet, as always: clumps of net and line, broken buoys, and bleach bottles. I've filled the *Alguita*'s dinghy as easily and quickly as I'd load a cart with groceries at the supermarket. It's been a dozen years since the 1997 voyage when I first saw plastic bits floating in the doldrums, and the debris cache keeps growing ever larger.

The windrow had aggregated and exposed what had been submerged and widely scattered flotsam in this seemingly clean spot far from the area I'd crossed in 1997. Our parachute anchor deployed, we'd lain at sea here for three days, waiting for a seaplane and the visitors it would bring. We'd chosen this site because it fell within the plane's range, but days of windy, choppy conditions were closing the safe-landing window. In any case, we doubted our guests would see much buoyant trash at these coordinates. For the last two days I'd motored the dinghy around *Alguita*, in ever-expanding circles, and found just a few plastic items. The third day, the winds calmed and the windrow serendipitously appeared, as if to prove there are no "fences" or demarcation lines for what had come to be known as the Great Pacific Garbage Patch. Here, where surface debris is widely dispersed, shearing wind over settled water had dredged up these cratefuls of plastic castoffs and fragments lurking just beneath the surface. Small, confused swells still prevented our visitors from safely landing. But as they banked and swooped back toward Honolulu, the conga line of debris made believers of them all.

I've learned to look with practiced eyes at the ocean's kinetic surface—called the neuston layer by marine scientists. The things I see teach me more than I care to know about the ways of mankind. Many of the objects here are still intact and once held food, drink, or product. Packaging. Much of what we're able to identify comes from Asia. Bigger stuff appears to be fishing flotsam—nets, buoys, line, and crates—evidence of a heedless industry. The scraps, the countless broken-down bits, will never yield their beginnings.

Now I know the plastic soup I saw on my first gyre crossing was inevitable. Since ancient times, humanity has tenaciously believed that the earth's oceans and waterways were there to make our waste disappear. We believed the ocean's capacity was limitless. Civilization's discharges have always been intimately connected with water. The word *sewer* finds its roots in the Anglo-Norman *sewere* ("water-course") and the Old French *sewiere* ("overflow channel for a fishpond"). The idea of water being the natural all-purpose cleansing agent has been around as long as we have.

The oceans seem boundless, like outer space. Stats: 67.7 percent of the earth's 198-million-square-mile surface—138.6 million square miles—is covered by salt water averaging two miles in depth. You can sail the seas for weeks and never catch sight of land. All of the earth's land area would fit into the Pacific Ocean alone. There are 353 quintillion gallons of water in the oceans. That's 353 billion billions, always in motion. Sylvia Earle, the celebrated oceanographer, says the planet should be called "Ocean," not Earth.

Yet in some ways I sense the oceans are shrinking. Science has shown how we've weakened them. We've seen the failure of the cod population to rebound on the Great Banks in the decade following a protective ban. Fossil fuel burning has rendered ocean waters 30 percent more acidic than fifty years ago, causing wide systemic stress. Long-banned synthetic chemicals and pesticides turn up in the tissues of apex predators like killer whales, dolphins, and seabirds. Half of the earth's 6.8 billion people live on or near coastlines. Most of the rest are near waterways that lead to the sea. Keep in mind that the oceans are downhill from nearly everyplace and everyone, and that plastics are nearly everywhere, used by nearly everyone, rich and poor alike. Ocean dumping may be deliberate or inadvertent; either way, it's nearly always preventable. Hurricanes and tsunamis can be horrifying natural forces that sweep debris from land into the sea—we saw this all too clearly in footage of the March 2011 tsunami in northern Japan—but we put the plastics there, within reach.

Now cut to a blustery day in November 2003. I'm leaning over a concrete rail, gazing down at an angry, rain-swollen urban river. I'm pondering how we can pull a net across this roaring thirty-mile-per-hour torrent without toppling the crane that works so well under normal conditions. Or ourselves becoming marine debris. The state of California has agreed it's worth knowing how much plastic waste is swept from the urban watershed into the oceans. With funding from a voter-approved clean water measure, the California State Water Resources Control Board is underwriting research we've proposed to make this determination. We'll take samples during normal dry weather, when the "rivers" are lazy streams. And we'll be poised to hit our sampling sites within twenty-four hours of measurable rainfall. That's when runoff will carry an extra-big dose of trash from a watershed populated by 13 million people and home to thousands of potentially polluting businesses. The rivers and creeks of Southern California used to meander. But days of torrential rain brought a catastrophic flood in 1938, killing more than one hundred people, destroying nearly six thousand homes, and bringing fast-growing Los Angeles to its knees. In response, the Army Corps of Engineers set about taming the Los Angeles and San Gabriel rivers with poured concrete. Four semitractor trailers could easily roar side by side down these channels where they're widest. Storm drains set into street gutters feed trash-laden runoff into the channels.

Using collection nets with varying mesh and mouth sizes, we

think we'll be catching a few surprises. We've decided to use different nets for different test sites rather than attempt to run one large net across the width of the rushing river. California has set limits for the total maximum daily loads (TMDLs) of substances that don't belong in clean water. That is, the amounts of chemical, metal, nutrient, or biological contaminants that a waterway can assimilate before beneficial uses—like swimming, fishing, and boating—are degraded. These allowable levels are supposed to approximate natural background levels of these substances. With trash, there's no natural background level, so the TMDL is set at zero. The state defines trash as any man-made particle 5 millimeters or larger. That means debris the size of buckshot, and smaller, is not, technically, trash. But we're afraid the small stuff is what's most potentially harmful to the marine food chain. We suspect and would like to prove that scads of so called microplastics collected from beaches and shore waters are spillage from the urban jungle. Once we've sourced them, maybe we can stop them.

Our work downstream from metropolitan L.A. strongly points to rivers as the primary source of plastics going into the ocean. More than beachgoers, more than past and present dumping from vessels, more than the fishing fleet. Rivers themselves are blameless, of course. We build our towns on rivers and coastal cities evolve where rivers empty into the sea. Rivers provide freshwater and food; they irrigate our crops; they can be harnessed for energy; and they float commercial boats and barges. Rivers have served us well, despite occasional floods, but we keep putting nasty stuff into them. Before the Industrial Revolution, it was mostly biological waste. Deadly cholera and typhoid outbreaks were regular results, though the connection between polluted water and infectious disease wasn't proved until the mid-nineteenth century. Indeed, the need to manage our excretions and discards has demanded much of man's ingenuity and often resulted in solutions with downsides.

Riversides proved irresistible sites for the Industrial Revolution's foundries, mills, factories, and slaughterhouses. Toxic discharge flowed unbridled, though not unnoticed. An early study conducted in 1839 by reformer Edwin Chadwick found that eight out of nine English laborers were dying of diseases caused by unsanitary conditions and polluted drinking water, not natural causes like old age or "violence." It was the same in industrialized areas of the United States, especially in poor immigrant enclaves. Well into the nineteenth century, most household waste, including excreta ("night soil"), was dumped in the streets, along with food scraps, and eagerly consumed by roaming pigs. It was a workable system. But in New York City, the pigs and the hardworking horses also produced half a ton of manure each day, making the city a nineteenth-century "nasal disaster." Ash, manure, and animal carcasses (fifteen thousand dead horses hauled from New York streets in 1880) were the headaches endured by pre-car, pre-electricity urbanites. By volume if not contents, per capita waste generation back then was comparable to our own today, about 1,500 pounds per person per year. But there were fewer people then, and the waste was natural and biodegradable, if noxious.

Efforts to provide sanitary relief proceeded in fits and starts, typically prompted by crisis: a cholera outbreak, harm to commerce, or offended citizens with clout. In New York City, river and ocean dumping had been routine. It finally came to a halt after the city's once-famous oyster beds had been killed and rich waterfront property owners threatened political consequences if animal carcasses and other refuse weren't stopped from landing on their beaches. Still, New York's new sewage system washed human waste into surrounding waters.

In 1899, Congress passed the Rivers and Harbors Act, barring dumping in navigable waterways, some, like the Chesapeake, had grown so choked with refuse that shipping was impeded. But the intent was to enable commerce, not to protect river ecosystems, a notion yet to be invented. Most larger cities were dumping organic waste into nearby waters well into the twentieth century. In 1918, a New York Academy of Medicine committee declared Manhattan "a body of land entirely surrounded by sewage." But again, protests centered on the human impact: sanitation, fouled surface water and beaches, bad smells. Even when fish were killed, as they were in Lake Michigan, reduction of the human food supply caused alarm, not degradation of the lake ecosystem. It wasn't until 1934 that Congress enacted a federal ban on near-shore dumping, but only of municipal garbage. The law exempted industrial and commercial concerns, paving the way for even worse pollution of rivers and shore waters when the postwar chemical age hit its stride.

In 1962, Rachel Carson's *Silent Spring* awakened readers to a new category of danger: man-made chemicals like the pesticide DDT. Suddenly, the postwar promise of "better living through chemistry" had an ominous undertone. Her book primed the public to react in a new way to the thirteenth fire to erupt on Ohio's oil-, chemical-, and sewage-polluted Cuyahoga River in 1969. The first twelve—starting in 1868—were recorded, but failed to ignite outrage.

The Cuyahoga is a minor waterway, only thirty miles long. But it flows through an industrial hotbed that includes Akron and empties into Lake Erie at Cleveland. In 1856, John D. Rockefeller sited his first oil refinery on the Cuyahoga. BF Goodrich, steel mills, and other dirty industries followed. Long devoid of life, the river was set on fire a last time by a spark thrown from a freight train on a rail overpass. The incident became a *Time* magazine cover story that was widely credited with assuring creation of the EPA in 1970 and passage of the Clean Water Act (CWA) in 1972. Then-president Richard Nixon signed the bills, despite industry's dire warnings that jobs would be lost and consumer prices driven up by higher production costs. They were wrong.

Even with a regulatory apparatus in place, loopholes, lax enforcement, and seamy politics allowed persistent violations and more pollution. New York and New Jersey kept dumping sewage sludge twelve miles offshore even after a mid-1960s study found that toxic metals and harmful bacteria were degrading the marine environment. It wasn't until 1988, by which time seven million tons of sludge had been deposited, that the EPA banned the practice. The solution: dumping sludge 100 miles out. As recently as 1987, more than 1,000 major industrial facilities and nearly 600 municipal sewage treatment plants discharged directly into estuaries or coastal waters around the country. In 1988, heavy rains on the eastern seaboard brought fresh calamity. Inundated beyond their capacity, sewage treatment plants overflowed, causing shoreline contamination with a whole new look. From New Jersey to New England, beaches were littered with plastic things, among them disposable syringes, diapers, and pink plastic tampon applicators—introduced in the sixties as a revolutionary breakthrough. These, along with toxic levels of bacteria and other contaminants, caused a rash of beach closures, cost local beach economies billions, and brought about the Ocean Dumping Ban Act of 1988. But neither this act nor the CWA brought an end to water pollution, though lawsuits brought by environmental groups to force enforcement were beginning to get results.

When an individual entity—say, a chemical plant—pollutes, it's called point source pollution. Water quality agencies are well equipped to deal with these polluters, and increased vigilance, monitoring, and enforcement have paid off. What's now getting into the oceans from rivers is mostly coming from non-point sources that can't be traced to individual polluters. It comes from beaches, roadsides, parks, cars, and stadiums. It's whipped from the backs of garbage trucks and from landfills before dozers cover the day's load with dirt or some "alternative daily cover." It overflows from garbage cans outside fast-food joints and dumpsters parked in alleys. Plastics of all shapes, sizes, and colors elude the waste stream and head oceanward, borne by breeze and water.

Now we're back at the Wardlow Road bridge, which spans the rain-swollen Los Angeles River. We determine that the smartest, safest way to drop our collection net is to attach weights at the bottom and floats at the top, then manually let it down into the center of the river. As the net rushes downriver, we'll drag it across the bottom and up the river's cement side. We'll wear life jackets and tether ourselves to a metal rail along the bike path. The water beneath the bridge is both turbulent and turbid, but chunks of debris break the surface as they tumble seaward. We position a lookout upstream with a walkie-talkie to alert us if something big—a log, say, or a sofa—is heading our way. We make out tree branches as well as plastic bottles, super-size cups, and shopping carts churning below our position. It looks like a sandbar on the cement bottom has developed. We drop the fine-mesh net into the center of the river and start tugging it toward the riverside. We're straining against the current and the net, which quickly grows heavy with catch. When we've hauled it to the river's edge, we hoist the net up the bank and empty its trashy contents into containers. We need to get three samples this way, as quickly as possible, and hurry the debris to our nearby lab for analysis.

Plastic is athletic. It scoots, flies, and swims. It travels without passport, crosses borders, and goes where it is, literally, an illegal alien. It has the endurance of a champ. It won't melt in water like paper or corrode like metal. Before bumping into the Garbage Patch, I helped with state monitoring of chemical, nutrient, and biological contaminants in shore waters off Southern California. These were stealth contaminants from untreated runoff and undertreated wastewater that were showing up in toxified coastal mollusks, marine dead zones, dying kelp beds, and waters unsafe for swimmers. Concern about visible matter—plastic bags, jugs, cups, straws, takeout containers, sandals, balls and balloons, and the broken-down bits—came later. After much of it had already colonized the oceans.

Nearly twenty years after his *Kon-Tiki* voyage, Thor Heyerdahl set out on a new quest. In May 1969, he launched a fifty-foot papyrus-reed raft, named *Ra I* for the Egyptian sun god, from a Moroccan port. He wanted to show that an ancient reed vessel could have crossed the Atlantic to the New World long before Columbus. As he and his crew struggled to keep the fragile craft afloat, they found themselves fending off tar clumps and watching man-made objects drift by. He wrote soon after, "Pollution observations were forced upon all expedition members by its grave nature . . ." The UN took note and urged Heyerdahl to take samples and keep a daily pollution log during a *Ra II* voyage planned for the following year. (*Ra I* wound up foundering just short of Barbados.) They logged tar clots from oil spills, but also plastic containers and rope, as well as metal cans and glass bottles. Heyerdahl told the UN:

> The present report has no other object than to call attention to the alarming fact that the Atlantic Ocean is becoming seriously polluted and that a continued indiscriminate use of the world's oceans as an international dumping ground for imperishable human refuse may have irreparable effects on the productivity and very survival of plant and animal species.

The report was filed with the UN's International Maritime Organization in 1970. Three years later the IMO ratified the International Convention for the Prevention of Pollution from Ships, known as MARPOL (MARine POLlution). In the slow-moving way of international treaties, especially those affecting trade, it would be another decade before the new law began to come into force. Though Heyerdahl's testimony helped, the chief impetus behind MARPOL was the disastrous 1967 wreck of the oil tanker *Torrey Canyon*. U.S.-built and -owned, but chartered by British Petroleum, the ship was the first supertanker to run aground. Its load—120,000 tons of Kuwaiti crude—inundated 120 miles of Cornwall coastline and 50 in France. An estimated 15,000 seabirds and untold numbers of shore creatures perished. So it was that MARPOL's first provision, Annex I, tackled oil transport. Annexes II, III, IV, and VI deal with, respectively, chemicals, packaged goods, sewage, and air pollution. Annex V, which bars marine garbage dumping, including plastics, came into force in 1988. By this time, plastic had conquered its rivals: glass, paper, and metal. Its production had already overtaken steel and the industry's growth rate surpassed that of all others. Until the last day of 1988, it was legal to dump plastics and any type of garbage in the oceans. Even now, compliance, technically speaking, is optional among non-signatory nations. In a seminal marine pollution review study published in 2002, New Zealand–based researcher José Guilherme Behrensdorf Derraik noted that MARPOL is still "widely ignored" and that ships are estimated to dump 6.5 million tons of plastic a year.

According to international law, individual countries have legal title to ocean waters within two hundred miles of their shores. Beyond these sovereign zones, the ocean belongs to no one and everyone, including you and me. The legal concept of "freedom of the seas" was one of the earliest tenets of international law. Hugo Grotius, a seventeenth-century Dutch jurist and poly-math, came up with the idea to legitimize Holland's freewheeling trade between Europe and the Spice Islands. His legacy lasted more than four hundred years, invoked when shipping and fish-ing rights came under dispute. Laws requiring marine conserva-tion measures are a twentieth-century phenomenon—as is the oil industry. Years before the *Torrey Canyon* disaster and MARPOL, oil contamination from ship spills had caused enough damage to local fisheries to prompt early stabs at international regulation. Even now, with anti-pollution rules firmly in place, the notion of "freedom" persists . . . if not always legally, then in the minds of many who ply the seas. It's a hard thing to surrender: the idea of a place left on earth where anything goes and no one will know the better.

If "developed" countries can't control ocean pollution, you can be sure the situation is worse in heavily populated countries where infrastructure and governance are weak. Countries like Bangladesh are MARPOL signatories but lack the means to enforce it—to stop dumping by foreign ships in their territorial waters. Bangladesh has at least banned the plastic bag. Even the least-developed places are knee-deep in the lubricant of global commerce: plastic. And this is often literal, because they lack adequate garbage management systems. Cheap goods are made of plastic. Shoes and gear are made of plastic; products and foods are contained in it. Trade newsletters tout growth in food processing, the great driver of plastic packaging in "emerging markets" like India and China. Sections of the Citarum River in Indonesia, a major waterway that provides drinking water for 80 percent of Java, are thickly covered by floating plastic trash. Fishermen are out of business—the fish are gone—and now reduced to picking through bobbing plastics in search of pieces that can be sold to recyclers. Children in India, the Philippines, and China do the same, immersing themselves in fouled waters as they troll for polyethylene.

On my first ocean crossing in 1961, aboard my father's forty-foot ketch, my deepest desire was to land a big fighting fish for a shipboard family feast. But even then, as a newly minted teenager, I was moved by the beautiful dominating expanse of blue, unbroken by anything. With chores done, I'd sit on the bow and watch the pristine surface of the ocean, waiting for something to happen. We'd jump up on the crest of a wave and watch the deck fall away beneath us. After a while, I might see the surface bumped by a silver silhouette, broken by a shark fin, or scrambled by a school of fish. A flying fish might launch itself across the wake, or, with luck, dolphins might draw alongside for a quick hello. Of inanimate objects, you might see a log or a Japanese glass fishing float, which you'd try to snag from the waves, later to hang on a porch beam. I finally got my fish, a mahimahi, and we never saw a speck of trash.

Now, anthropogenic debris—man-made trash, 80 to 90 percent of it plastic—has broken the reverie of pristine perfection that is the ocean's essence. It's become her most common surface feature. Now, depending where you are in the ocean, you'll see dozens of balloons, buoys, and bottles bobbing by before ever catching sight of a leaping tuna. Trash has superseded the natural ocean sights, stamping a permanent plastic footprint on the ocean's surface. In 1951, Rachel Carson wrote in *The Sea Around Us*: "The face of the sea is always changing. Crossed by colors, lights, and moving shadows, sparkling in the sun, mysterious in the twilight. . . ." I wonder what she'd have written about the sea's now plastic-pocked face. In 1951, *The Sea Around Us* and *Kon-Tiki* were both top-ten nonfiction best sellers. This was also the year Phillips Petroleum research chemists J. Paul Hogan and Robert L. Banks developed new catalyzing methods enabling commercial mass production of HDPE—high-density polyethylene—and polypropylene. The plastic plague loomed.

In 1975, the National Academy of Sciences estimated that 14 billion pounds of garbage were being dumped into the ocean every year from boats and ships, a third of it by U.S. vessels. Aircraft carriers with crews of 6,000 sailors generate over three million pounds of trash during a six-month tour. The Navy self-reported in the 1980s that plastics probably accounted for 12 percent of all waste generated on board. That means more than 300,000 pounds of plastics dumped per tour by a single vessel before MARPOL V kicked in. By its own admission, the Navy has added more than 4.5 million pounds of plastic to the world's oceans. In degraded form, most of it would still be there. (The U.S. Navy may be the worst ocean polluter the world has ever known, having secretly dumped, by its own account, 64 million pounds of nerve and mustard agents into the sea, along with

400,000 chemical-filled bombs, land mines, and rockets, and more than 500 tons of radioactive waste—either tossed overboard or packed into the holds of scuttled vessels.)

In 1982, a federally funded study published in *Marine Pollution Bulletin* estimated that 639,000 plastic containers were being dumped daily from merchant ships. Though they're the most massive ships in the ocean, supertankers and container ships carry small crews of ten to twenty. But container ships are prone to losing cargo-filled containers in heavy seas, up to 10,000 a year in the 1990s, before loading technology improved. These accidental losses are not required to be reported. By contrast, cruise ships, of which about 300 are in operation around the world, hold 3,000 to 5,000 passengers and crew and service more than 14 million passengers yearly. The typical cruise ship generates about 50 tons of solid waste, including plastics, during a one-week cruise. After a spate of MARPOL Annex V violations in the 1990s, cruise ships cleaned up their act. Now ocean liners incinerate, compact, grind, or recycle most of their collected waste. Some naval ships are equipped with plastic densification technology, machines that compress waste plastics into pizza-shaped discs for storage and later disposal. But do we really think dumping has stopped? The key to knowing would be technology to date plastics captured from mid-ocean. A method for doing so remains elusive.

No one really believes MARPOL has put an end to ocean dumping, though it has helped. Because of what I do, people approach me with their stories. A former sailor told me ocean dumping was happening on his Navy ship in the nineties. A merchant marine sent an anonymous message that he sees ocean dumping daily from container ships. He's afraid to speak out for fear of losing his job. I was amazed by an anecdote in the book *A Captain's Duty* by Richard Phillips, the merchant skipper who, in 2009, ably managed a harrowing Somali pirate hostage ordeal. The Navy had become involved in his rescue. It was dark. Everyone was jumpy. Phillips was a hostage on the pirate skiff. Suddenly, he and the pirates heard plopping sounds and saw black blobs floating by. The pirates assumed it was a stealth maneuver and got on the radio, shouting, "No action! No action!" — but Phillips knew it was just garbage. He remarked: "Merchant ships can't dispose of plastics on the ocean but the Navy can. The Navy confirmed it. They told the pirates it was just garbage floating away." If the Navy is truly exempt, it's a rule I've never come across, and a very bad one. Hearsay reports abound of plastic cups, balloons, streamers, and stuffed garbage bags appearing in the wakes of Navy cruise ships.

Larger U.S. ships are required to file reports with the U.S. Coast Guard at the beginning and end of each voyage, accounting for material on board when they leave and what comes back to port. Manpower limitations make verification rare. So, are the laws helping? There's no research that can answer that question, though our studies since my first voyage through the North Pacific High suggest not enough—not even close. Old habits die hard, and deepwater enforcement of anti-dumping laws is a pipe dream. In most ports, ships looking to offload their trash need to pay for disposal, a disincentive if ever there was one. Not all ports are even set up for collecting shipboard trash, although the port of Rotterdam's "Any Waste Any Time" initiative is a welcome exception and appears to be catching on in other European ports. Even under MARPOL Annex V, "accidental" loss is okay. So, do we have any idea about the size of the oceans' plastic burden? Do we know who the worst dumpers are? Both answers are anyone's and everyone's guess. The United Nations Environmental Program (UNEP) has opted out of the guessing game after floating various widely reported estimates. The most recent, from 2009, put the ocean's plastic burden at about 615 million metric tons. Other now-defunct estimates proferred by UNEP and its Joint Group of Experts on the Scientific Aspects of Marine Environmental Protection (GESAMP) had five million pieces of plastic entering the oceans each day and 13,000 pieces of plastic residing in each square kilometer of ocean. These figures were helpfully alarming, but they could not be scientifically supported. UNEP's 2011 edition of its annual *Year Book on Emerging Issues in Our Global Environment* includes for the first time the dedicated section "Plastic Debris in the Ocean." The issues are well limned in this report, and credit is given to Algalita research, but quantification estimates are conspicuously absent. The report concedes, "It is difficult to quantify the amounts and sources of plastic and other types of debris entering the ocean." And it further ventures that "a comprehensive set of environmental indicators for use in assessments has been lacking." In other words, we are far better at describing the issue than measuring it.

The Plastic Age has sneaked up on us almost imperceptibly, and noticing waste plastic in the ocean forced on us the realization that something was changing. Plastic trash in such a pristine environment—as far as you can get from places where it's made and used—was an illuminator. Perhaps for a while we weren't as bothered as we might have been, because we still thought plastic material was inert and benign, an eyesore that couldn't do much harm. Now we know better. But even before I learned about the varieties of harm done by plastic debris, and of its toxic potential, seeing it out there where it didn't belong seemed profoundly wrong.

Back at our urban river project, we truck our three days' worth of samples to the Sea Lab and begin analyzing the samples. It's a painstaking business, all the more painful because the results are so shocking. During three days of sampling, extrapolating from our "snapshot" grabs, we calculate the Los Angeles and San Gabriel rivers discharged no less than 2.3 billion pieces of plastic into the ocean, total weight roughly thirty tons. We now respectfully disbelieve the UN's good-faith estimate that has about five million pieces of plastic moving from land to oceans each day. Do we need to know how much plastic is floating around in our oceans and do we need to know if it comes from land or sea sources? Isn't any too much? Yes and yes. But we need baseline figures to tell us whether anti-dumping laws are working. This has been my work. We also need to reliably source the debris so efforts to staunch the flow can be targeted and do what they're supposed to do: keep plastics out of oceans. We need to draw a line in the sand and say to plastic, "This far and no further."

CHAPTER 5

THE PLASTIC SEA AROUND US

WE'RE DOWN TO THE WIRE. We're scheduled to leave port on this first official expedition to the gyre on August 15, 1999, and already it's late July. As per Curt Ebbesmeyer's advice, my main focus is coming up with a net system that will snag tiny broken-down bits of plastic as well as bigger chunks, thereby achieving a holistic portrait of the gyre's contents.

First, I look at various types of hoop nets used to collect plankton. Then I decide it might be smart to consult with one of the players in the Bight '98 study, Chuck Mitchell of MBC Applied Environmental Sciences. He invites me to his Costa Mesa facility to check out his trawling equipment. I've finally come to the right place. Chuck's on-site warehouse is crammed with all type and manner of oceanographic sampling gear, things I've never seen, let alone heard of. We talk about my conversations with Ebbesmeyer, how we'll be trawling for the smallest debris out there, microplastic shards, many of them too small to spot with the naked eye. Chuck tells me a typical plankton hoop net is not what I want and shows me a device he insists will do the job better. He calls it a manta trawl, and says it's normally used to skim fish eggs and larvae from the neuston layer. For purposes of scientific sampling, he says, this net offers superior catching ability at the "air and sea interface." Even better, the net will make it easier to quantify the catch per unit of water because it's designed to consistently present the same opening to the neuston. Consistency and control are a researcher's best friends. The manta trawl has a wide, hooded, thirty-six-by-six-inch mouth and two large stabilizing wings resembling those of a manta ray. The six-foot-long net end looks like a windsock, and the collection bag is very fine 0.33 millimeter mesh, much like cheesecloth. The collection bag at the "cod" end detaches so the sample can be decanted into a storage jar.

It's a first. A manta net has never been deployed as a plastics catcher. We'll still run the half-inch mesh trawl between the two hulls, and we'll still augment with the subsurface "otter trawl"—*otter* referring not to the cuddly looking marine mammal but to sailor lingo for "outer" boards that hold the net open. This one normally has a 5-millimeter-mesh collection net at the cod end, peppercorn-sized, small enough to catch about half of James Marcus's Hawaii beach sample. So we modify it by sewing in 0.33-millimeter-mesh cloth to make it comparable to the manta. We're also equipped with a side-scan sonar meant to detect larger debris chunks suspended beneath the sea surface. These we intend to haul out for closer inspection, possibly to collect for a trip back to their terrestrial origins.

Chuck is intrigued by our project and offers the net on loan. What a guy. Let it be said, part of Algalita's credo is to do what's necessary to get the job done, but do it frugally, with volunteer crews and donated supplies whenever they're offered. The entire expedition is budgeted at $3,350, with diesel fuel the largest single expense. Not bad for a three-week voyage: a week sailing to the gyre, another surveying, and a final week's voyage back to port.

A cheery group of about twenty friends, family, colleagues, and local press gathers on the dock to send us off on a sunny Sunday mid-August morning. We sail the hundred miles up to Santa Barbara, there to top off the fuel supply and take delivery of several crates of organic vegetables. These are from future crew member Chris Thompson, a friend and organic grower for John Givens Farms, marketed under the Something Good label. This connection helps us get more local press in Santa Barbara. I've gone back to these stories and read with keen interest what we and others said about this voyage, before anyone had the slightest notion of what we'd actually find out there. I say to the *Long Beach Press-Telegram*: "People need to know that we are filling the ocean up with plastic stuff. . . . Years ago, I never really noticed that kind of debris. Now, it is the most distinguishing thing out there." I say our goal is to make the oceans "swimmable," for people and the oceans' denizens. We talk about Ebbesmeyer's predicted toys, shoes, and hockey gloves, and the recent storm that pitched four hundred cargo containers into the north-central Pacific. Hundreds of the toys and pieces of athletic gear have washed up on West Coast beaches in the last decade. But tens of thousands were lost in the spills and are unaccounted for. According to OSCURS, the computerized ocean current model, a good portion of them should have trailed into the Garbage Patch.

What I've seen and learned thus far points to a situation deserving much attention, perhaps requiring a paradigm shift. This will be science, but if our hypothesis is borne out, I promise myself it will not be vacuum packed. We're prepared for anything, and hoping for a few surprises.

It's a rough beat out past Point Conception, the natural demarcation point between southern and central California. This massive promontory juts thirty miles out to sea just north of Santa Barbara, the northern boundary of the Southern California Bight. The crew, only one of whom (my neighbor Mike Baker) has sailed the high seas, is shaken down (and up) by high winds and seas. *Alguita*'s first-aid station is well stocked with seasickness remedies, the best of which are transdermal scopolamine patches, derived from the belladonna plant family, and tried-and-true Dramamine tablets. I dispense them as needed. In our first surprise, this maiden day under sail is spent dodging missiles fired by U.S. Navy warships practicing in the area. The Navy is chasing fishing vessels out of the firing zone. It occurs to us that with the pros flushed out, the fishing might be good. We put out a line and within minutes hook something substantial. Ten strenuous minutes later we land a twenty-four-pound albacore that was, like us, heading westward to the gyre. After several days of choppy upwind sailing, we encounter the first calm day of the trip, about five hundred miles out. We decide to do a practice trawl. We're only halfway to the central gyre, so we don't expect to catch anything except plankton. But we do. A lot of plastic chips and specks. Though it's not part of the official sample pattern, we'll note this trial trawl in the log and save its contents for later inclusion in our database. It's a surprising and sobering find, one that ratchets up our collective sense of purpose and indignation.

Crew member and artist Steve McLeod impresses us all with his sharp beachcomber's eye and iron concentration. With the patience of a Himalayan yogi, he stands on the bow for hours at a time, drawing each visible floater within eyeshot on a transect chart. This gives us a good record of the debris that lies beyond the reach of our nets, or is too big to land, or doesn't quite warrant dropping the sails and deploying the dinghy. McLeod has played a crucial role in Ebbesmeyer's flotsam studies, being the one who ran a clearinghouse for matching recovered Nike Air Jordans lost in the 1990 *Hansa Carrier* container spill. And he's managed the rare feat of eking out a living as an artist. His paintings are mostly seascapes, admired for their moody, luminous quality. He has a special talent for transforming beachcombed man-made flotsam and natural items into works of art. I was honored when he presented me with a walking stick crafted from bull kelp, cured, dried, and with a moonstone topknot. My cane is as sturdy as any made of birchwood, and thus far no visitor has managed to guess the material it's made of. I call it "nature's plastic." Steve's artistry will serve us well down the line, when he donates several pieces of flotsam art to an Algalita fund-raiser and actor Ed Asner ably auctions them to the highest bidder.

McLeod also has a flair for metaphor. He says the countless plastic chips suspended throughout the ocean's top layer remind him of the star-spangled night sky during his night watch shifts—cloudless and clear out here, far from civilization's competing illumination and the coastal fog of his Oregon coast. Many of these brightly colored fragments end up as what Steve ironically terms *decorations* along the surf line of the beaches he combs. As bad as if a majestic forest were to be hung with gaudy, tasteless

Christmas ornaments, he says. The sadness of it hits us all as we listen to Steve. Our children may never know beaches free of plastic sand or a sea devoid of plastic gobs and flecks.

Eight days out of port, we're where we need to be to start sampling, near the central cell of the subtropical high. The wind drops to below ten knots and we deploy the manta trawl for our first official sample at the eastern edge of the central gyre, about eight hundred miles from land. The net skims along the ocean surface, funneling debris into its very fine mesh filter. It's a smart piece of equipment and we're impressed by its performance. After trawling for 3.5 miles, we haul the net in to examine our catch. We carefully rinse down the part of our haul that clings to the manta's tapering sides. At the end of the net is the mesh collection bag, about the size of a quart jar. It's clamped around a black plastic tube, which in turn is clamped inside the open end of the net. We nervously detach the collection bag and cluster in for a peek. I don't know that we'd actually expected to see what was there, despite all the signs. The plankton presents as a gelatinous mass. It is laced throughout with plastic chips, like tutti-frutti in a melting gelato. Plastic bits appear to rival even the volume of plankton, but we won't know for sure until later, after the samples have undergone lab analysis.

"Plankton," by the way, is rooted in the Greek word *plank-tos*, meaning errant or wandering. *Plankton* is a catchall term that includes myriad plant (phytoplankton) and animal (zooplankton) organisms that range from the microscopic to the visible. Some mobile planktonic organisms engage in a daily migration from the depths to the surface of the water column, but these basic life-forms mostly go where ocean currents and breezes take them. Much like plastic debris.

Our resident birder, Robb Hamilton, has been adding pelagic

species to his list—the list all serious birders keep to record and describe each new bird species they sight. In fact, he seems to have attracted an avian companion, a handsome black-footed albatross, which swoops by for a visit several days running. An acrobatic red-footed booby entertains us on the eleventh day, a day that ends on a sad note. Now one thousand miles from the California coast and halfway through our sample trawls, an exhausted young Baird's sandpiper ditches near our boat. This is a species whose late-summer migration normally takes it straight down the center of North America. Using a net, Robb manages to fish the poor bird from the water and lay him gently in a box he's lined with padding. He calls for sugar water, prompting a rush to the galley. We also produce a bit of fresh mussel that happened to be hitchhiking on a stray fishing float we snagged. But our attempts to revive the bird are fruitless. The sandpiper goes limp. We decide to freeze his body to donate to a museum. His final resting place turns out to be the archives at the Los Angeles Museum of Natural History. Had we known then what we know now, we might have chosen instead to arrange a necropsy, to see if plastic had found its way into the bird's diet.

Meanwhile, the otter and the net suspended between the hulls are *not* grabbing macroplastics by the bushel as expected. I'm especially disappointed by the net strung between the stern swim steps. It's a good thing we're not relying on it exclusively, per the original plan. We use the otter trawl to do subsurface sampling of lower pelagic layers, and what it mainly traps are short monofilament fibers coated with algae. The little bit of weight from the algae evidently causes the scraps to drift down to deeper levels in the water column. Lesson learned: the plastics out here primarily populate surface water.

Our sample design calls for random trawl lengths and random

distances between trawls. Molly Leecaster has provided us with guidelines. On the way to our next manta deployment, we again experience the calming of wind and sea state—the doldrums—that tell us we're near the center of the eastern eye of the gyre. Given these perfect conditions, we scramble to deploy three nets at once: the manta, the otter, and the transom trawl with its half-inch mesh, to catch larger items (e.g., bottle caps, net and rope scraps, plastic bags) at the surface. They are deployed in such a way that none will impinge on another's potential catch. We need to rev up the engines to maintain our trawling speed between 1.5 and 3 knots. At the end, we find that each net has grabbed trash, but the otter trawl, set at a depth just over thirty feet (ten meters), gets the least. This is no surprise. Buoyant plastic types will waft around in the upper layer, rising nearly to the surface in calm seas, while, as we saw with the algae-coated line, denser and organism-encrusted plastics sink to lower levels in the water column. Once again, the manta catches the most.

The samples never cease to astound us. Each seems to have its own character, and in each the crew and I find the manta's "stomach contents" to be amazing and disturbing. Not one is free of plastic. Indeed, the manta, as billed, provides an effective, reproducible method for sampling anything and everything that floats on and near the ocean's surface, be it plastic, plankton, plankton-eating filter feeders, or small fish, as well as odd flecks of paint and tar we later extract from the sample clumps.

On our last day sampling, we spot a large, mostly submerged piece of flotsam that seems to peek at us from just above the surface. Crew members toss a marker buoy and then jump in the dinghy to attempt recovery. This is one-stop shopping at its worst, and it teaches me a lesson about the ocean. It knits and it weaves. Distant, disparate, yet in same way similar objects find each other in the middle of millions of square miles of seemingly empty ocean, and the ocean stitches them together, making a grotesque whole. We dive around this strange mass, avoiding the undulating tendrils of rope and net and deciding it could make a convincing sci-fi monster. Once we've grappled it on board with assorted hooks and lines, we itemize the contents. It's mostly composed of variegated polypropylene fishing nets and lines, so we dub the beast "Poly P." Gamely, and agile as a seal, John Barth, the retired head lifeguard, slips into the water to wrangle a fully inflated truck tire and rim to *Alguita*'s hull, where we can get a purchase on it. It's a nasty piece of work, clearly a longtime gyre resident, bearded by gooseneck barnacles on the rubber tire and coated by algae on the steel rim.

We collect an unholy mass of trash. Besides heaps of netting and rope, we catch a chemical drum, an embrittled bleach bottle, several Japanese net floats, a foam sheet with shoe soles cut out, and a Dairyland sour cream container. We've recorded each as to recovery position and weight, first "as caught," barnacles and all, and a second time scraped clean. This tells us the quantity of "fouling organisms," otherwise known as hitchhikers. Fouling "communities" are typically associated with contaminated hulls of boats that travel from port to port, seeding local waters with alien and often destructive invasive species. In the world of marine science, they are called sessile organisms, *sessile* meaning attached. In their earliest, larval stages they must "recruit" an object to anchor on or else perish. These organisms seem to love plastic flotsam. If their migratory homes stay whole, it's anyone's guess which distant habitats will greet unwelcome colonists. We also note that these creatures coat the plastic in a way that appears to shield the object from the sun's penetrating UV rays, at least on the surface below water. It's a new kind of symbiosis, with the biological ballast slowing embrittlement of the plastic by keeping one side down.

After a few days of trawling, spotting, retrieving, and cataloging debris, we begin to have a feeling for what we're truly looking at in the Mid-Pacific Gyre—something worse than we'd expected. On land, it's soothing to think that all those bottles and wrappers, all that cheap plastic stuff we handle every day, winds up in a landfill, safely sequestered from polite society. But here in mid-ocean we're finding hordes of escapees from imperfect collection systems (in some places, none at all) and seeing flaws in hard-to-enforce international marine pollution laws. All this wayward plastic dreck is beginning to look like civilization's dirty little secret. Try as we may to control it, to hide it, to manage it—it mocks us and goes where it doesn't belong.

We spend a good deal of time in the water, and it's not exactly like swimming with the dolphins. Part of our protocol after landing a trawled sample is to jump into the ocean with fins, snorkel, and mask to visually confirm the presence of plastic drifters in the water column. A typical plunge—fifteen minutes in 74-degree Fahrenheit water—yields a handful of drifting plastic fibers and visuals of tiny plastic specks floating by on the current alongside barely visible organisms that populate the area. The plastics do indeed seem to mimic the plankton—bad news for plankton eaters. Alas, on this voyage, we have no underwater camera with which to record our finds, a situation I remedy on future expeditions.

We begin to see larger planktonic creatures known as salps. These are basically feeding tubes made of clear jelly, though,

surprisingly, they're members of the phylum Chordata, meaning
they're vertebrates—albeit in primitive form—like us. In southern oceans, salps colonize in rows and sheets that resemble fantastic honeycombs. We've seen reports of salp colonies turning
square miles of the ocean surface into shimmering fields of
jelly. Now we've learned they play a significant and beneficial
role in the carbon cycle. These tunicates, as they're also known,
pump their way through the neuston layer, indiscriminately
inhaling algae, diatoms, phytoplankton, small zooplankton, miscellaneous bits of food left by other creatures, even bacteria, and
then excrete the digested remains. It's odd to see this excreta
coming from a nearly transparent organism. Many of the salps
we encounter sport plastics, inside and out, little colorful chips
embedded in clear tissue. They still do their thing, but they're
festooned. An especially large specimen, the size of a cardboard
toilet paper roll, winds up in the manta net, speckled with plastic
fragments. It's hard not to wonder if plastic is affecting these
exotic organisms in ways unknown. We wonder as well about
food-chain higher-ups that eat them. Maybe the small stuff just
passes through the creatures that gulp it down. It would seem
that way. A study of ingestion out here would be interesting,
I think.

During the expedition, we use our laptop and printer to pub-
lish an onboard newsletter that will be distributed by Algalita
staff soon after our return. In reading my old dispatches about
this voyage, I'm slightly chagrined seeing how I then called all
plastic bits by the term *nurdle*. It should have been reserved only
for those preproduction pellets, which happen to be scarce in the
pelagic habitat. Here's a sample: "Ladies and gentlemen, millions
of square miles of the Pacific Ocean—the surface waters com-

prising its most remote and pristine center—have become nurdle soup! Dive in anywhere in the High and watch nurdles float by. Sail through it and watch nurdles float by. Drag a net through it and catch them by the thousands. Over the 50 odd years since its introduction, decomposed plastic has established itself in the Pacific as a ubiquitous, totally non-nutritive component of the plankton." I find I also referred to plastics as being "biologically inert," which I now know to be false. We're soon enough to learn they are in varying degrees bioactive. The prevailing assumptions at the time now seem naïve. But the convictions remain unchanged.

The term *non-biodegradable* is often applied to plastics, meaning they defy digestion by living organisms. Now research has shown that certain microorganisms very slowly biodegrade plastics in certain conditions. The sessile creatures homesteading on plastic chunks in the gyre seem to cause pitting, but we've yet to learn if it's evidence of mechanical damage or biological digestion. But what I call "enviro-degradation" operates more swiftly. The plastic shards we found in the gyre were mostly broken down by exposure. Plastics have a curious in-between form of permanence. Relative to other materials like glass, steel, and rock, plastic objects lose their physical integrity quickly, especially when exposed to sunlight and mechanical and oxidative stress. But they are far stronger than biological tissue. When we die, so do our cells. We give off methane and hydrogen sulfide gases. Our physical matter undergoes a complex transformation—with many organisms eager to help—culminating in dust. Plastics, on the other hand, break into smaller and smaller pieces, but their molecules are still little polymer fibers that may exist as infinitesimal entities for centuries, perhaps millennia. Because plastics are relatively new to this earth, it's simply too soon to know how long they will persist, and what the consequences might be.

Recall that I'd thought James Marcus's plastic bits from Waimanalo Beach had been mechanically ground prior to shipping to a reprocessor. Before embarking on this voyage, I studied up on plastic decay. Ebbesmeyer recommended I read studies by Dr. Anthony Andrady, the leading researcher in this field. Out in the gyre, I realize I'm seeing in situ what his studies uncovered. Like vampires, plastics don't react well to sunlight. UV radiation embrittles plastic by de-linking the polymer chains, which is what's happening when a vinyl car dashboard cracks after years in the sun. Seawater begins to leach out additive chemicals from the weakened plastic, chemicals that once enhanced its durability and flexibility. Andrady is in fact the expert who, after years of research, declared that all estimates about plastic's persistence are only guesses at this point.

I had questioned Ebbesmeyer's belief that plastic things break down in the gyre simply from UV exposure. After this trip, we wind up agreeing that a number of scenarios are playing out. Some of the gyre plastics probably began degrading on land, with an assist from pulverizing shoreline waves. But I have another theory. Given the bite marks on many of these plastics, I'm convinced a goodly number of these fragments were nibbled off larger plastic objects by hungry fish and passed as excreta.

Ebbesmeyer made a provocative calculation in one of his letters. By his estimate, thirty liter-sized PET soda bottles, each breaking down to 12,500 granules the size of the particles in James Marcus's Waimanalo Beach sample, can provide a plastic bit for each of the 372,000 miles of coastline on earth. What we're seeing in the gyre is a half century of continuous, cumulative plastic deposition. It's a marine museum of the history of plastic trash. But we've yet to learn how to extract the information we really need from these fragments, namely their sources. Controlled studies reveal the mechanics of plastic decay, but a jar of plastic bits trawled from the gyre tells you very little. The variables are too many. Plastics don't yet have a DNA code we can read, and the same plastic types are used throughout the world. You could say the gyre is a plastics melting pot. What we know for certain is that plastic trash is gathering out here, and it rivals some of the natural constituents of the sea surface.

The gyre's sunny skies and warm, tranquil waters afford us five busy days of sampling followed by a blissful day of R&R. We bask in the sun and plunge into the warm ocean for some recreational snorkeling, still swarmed by plastic scraps. A few tropical birds lazily glide overhead. A warm front barrels through the high after our day of rest and blows away any chance of further sampling. We hop on this weather freight train and begin our trek back to Santa Barbara. No need to crank up the engines. Owing to these favorable winds, we arrive two days ahead of schedule with fuel to spare. It's the exact opposite of the 1997 return voyage from Hawaii, which serves to illustrate the disruptive power of El Niño events. Afterward, more press. Artist/beachcomber McLeod tells the Santa Barbara newspaper: "There were areas where it was just coming by in a steady stream. That's distressing."

Back in Long Beach, my top priority is to contact Ebbesmeyer, whose excitement about the expedition equaled ours. I call him within a day of docking and he says, "Come on up." I think: why not? There's no time to waste, because I've discovered a mature oyster attached to a broken buoy. Ebbesmeyer says if I can get it up to him in good enough shape, he knows someone who can identify it and determine its place of origin. It's *CSI*-style debris forensics.

My work truck isn't suitable for this long haul, so I ask to borrow my mother's '91 "matron beige" Cadillac Coupe de Ville. I leave on September 22, less than two weeks after our return. I've lined the backseat and trunk with plastic tarp and loaded in heaps of debris. Some may be traceable to one of Ebbesmeyer's container spills.

This is not a leisure trip. I head up Interstate 5, through the flat agricultural midsection of California that might as well be the Midwest. The road begins to climb in the vicinity of Mount Shasta, just shy of the Oregon border. The Caddy starts overheating. I'm reminded that this is a GM car, and GM is credited with inventing planned obsolescence. I've got some gallon polyethylene jugs and fill them with water, pulling over every fifteen minutes or so to quench the overheating engine. I finally reach a town large enough to have a Cadillac dealership. The service department tells me it's cheaper to replace the whole engine than just the head, which is warped aluminum and beyond repair. I leave my mother's Caddy behind and continue on in a loaner while the mechanics swap out the old engine for a "new" reconditioned one.

I reach Seattle two days after leaving Long Beach, a tiring overland voyage. But Seattle never fails to amaze me, with majestic Mount Rainier looming close behind and the sparkling sweep of the sound. I pull up near Ebbesmeyer's wood-frame home with cherry trees in front. He comes out to greet me and suggests I arrange the debris on his front lawn. I present him with the well-traveled oyster on its broken plastic buoy. We've been through a lot together, that bivalve and I.

As far as he can tell, none of the stuff we'd dug out of the gyre is from any container spill he knows of. The volleyball is too weathered to be from the "meteorological bomb" of a year earlier, in October 1998, and much of the rest came from commercial fishing activities. No rubber duckies, frogs, turtles, or beavers, no hockey gloves, no Air Jordans. The eastern gyre is vast beyond description, and the odds that we'd recover specific items were, after all, needle-in-a-haystack low. A land-based analogy would be a circular highway running through two Texases. You would drive the highway in a golf cart, looking for, say, $20 bills that might have fallen out of Texans' pockets. But the open ocean is not a fixed highway and has no landmarks to follow. And remember, roughly half our days out in the gyre were spent in darkness. A mystery of the gyre is this: if you're looking for some specific thing out there, even if it's predicted to be there by a computer model, you probably won't find it. But you will find many other misplaced things that possess the strange ability to inspire both excitement and dread. Like a horror show.

This is when I discover that Ebbesmeyer is quite a different cat from what I'd assumed. At no point has it occurred to me that Ebbesmeyer, a Ph.D., is anything but a university professor. I learn otherwise when I ask to see his lab and he takes me down to his basement, a flotsam catacomb. He shows me his collection, much of it arranged in tackle boxes. I hoped to see a typical marine science lab, facing as I am the task of processing and analyzing our rigorously collected gyre samples. Ebbesmeyer tells me his specialties are offshore drilling platform construction and safety and oil spill management. He's worked for oil and consulting companies with assignments all over the world. Now he occasionally consults. His great, consuming love is beachcombing and hanging out with beachcombers at local meetings and annual gatherings where flotsam fanciers from everywhere compare, share, sell, and swap their finds. Like most beachcombers, Ebbesmeyer is a connoisseur of Japanese glass fishing floats and has acquired a fine collection.

My host decides to call a fellow flotsam expert, Steve Ignell. I know his name from the marine debris study that he, Robert Day, and David Shaw—the Alaska-based scientists—conducted in the 1980s, in hopes we can sync sample analysis protocols and swap information. Ebbesmeyer has revisited their study and is convinced the ocean's plastic content is growing despite MARPOL Annex V. The prospect of comparing our results with the earlier study has him at the edge of his seat. I listen as he speaks to Ignell and then begins to look a little crestfallen. He hangs up the phone and tells me Ignell said, "We already did that." End of story. They found and measured plastic in the Pacific from 1984 to 1988 and published their findings in 1990. More than a decade later, we investigated a significant area outside the scope of their study, and we had better equipment. I'm not quite getting the lack of interest. But we move on, sitting at Ebbesmeyer's dining table and plotting other ways to figure out where all this gyre stuff is coming from and where it's going and how to make it stop.

On the drive back home, I experience an upwelling of frustration. Ignell, Day, and Shaw had found a notable plastic presence in the Japan Sea and the North Pacific more than a decade earlier, and nothing happened. Of all the people in the world, I—who created a marine research foundation and studied water quality and cleaned up coastal wetlands and was the Coast Guard's go-to guy for plastic confetti found on a beach in Hawaii—I should have known that other people already knew the oceans were filling up with plastic. But I didn't. Somehow the word hadn't gotten out. And this is when I realize it will take nothing less than a crusade to stop the unspooling of this man-made horror show in the Pacific, and possibly in all the earth's oceans. A campaign rooted in science but driven by passion. And I've yet to learn the worst.

CHAPTER 6

THE INVENTION OF
THROWAWAY LIVING

THE WINDS OF CHANGE blew through our kitchen in the late 1950s. The milkman was the harbinger. In his crisp white uniform and sporty cap, Paul would show up each week with a steel carrier jingling with glass quart bottles of milk and a lamb bone for our dachshund, Pipperly von Brunigswagen. We all liked him, especially Pipperly, normally a tough customer. Paul would reload his carrier with rinsed bottles from the last time and be on his way. Back at the dairy, about twenty miles from our house, the bottles got sterilized, refilled, and loaded back on the truck to deliver again. Imagine that. With benefit of hindsight, this model looks not boringly normal but innovative and green.

One day the milkman arrived with half gallons of milk in printed paperboard cartons. That seemed a little odd, perhaps a little excitingly new. We adapted. Did Paul know these cartons spelled his doom—that he was already an endangered species, soon to be swept aside along with his reusable glass bottles, local dairies, and hormone-free milk with cream on top?

Early milk cartons were waterproofed with paraffin wax, a petroleum product. Schoolchildren scraped it off their little milk and juice cartons with their fingernails. By the 1970s, polyethylene-laminated paperboard dominated. No more wax worms. High-density polyethylene gallon jugs arrived in 1964, sealing the milkman's fate. Except in rare pockets of the country, some with drive-through "milk barns," milk became a supermarket product, purveyed from afar in containers meant to be trashed. One after another, local grocers, bakers, and butchers quietly shuttered their businesses as sparkly clean chain supermarkets erupted from the middle of vast parking lots. At the supermarket, milk in cartons, bread in plastic bags, and precut meat on Styrofoam trays cost less and somehow seemed more hygienic.

If there is a birthday for the new age of disposability, it would be August 1, 1955. In that year TV ownership hit critical mass— 65 percent of households—and "consumerism" was taking off like a dragster. On that day, an article appeared in *Life*, the nation's leading magazine, with a circulation of 12 million and a readership far greater. Its title was "Throwaway Living: Dozens of Disposable Housewares Eliminate the Chore of Cleaning Up." An accompanying, now iconic photograph by Peter Stackpole shows "objects flying through the air": a jumble of airborne aluminum pie tins and sectional plates, paper napkins, paper cups ("for beer and highballs"), utensils—even a disposable diaper, given credit for fueling the postwar baby boom. Standing beneath the expendable array are a woman in a "housedress," a little girl in a smocked dress and Mary Janes, and a "hep cat" in white T-shirt and rumpled dungarees. The unlikely trio gazes upward from behind a wire mesh trash basket from which, or into which, the objects fly. No matter the trajectory. In this photo we clearly see the early signs of a trash avalanche that will have sanitation engineers scrambling. Worse still is what it means for the oceans.

Tellingly, the text makes no mention of plastics. The flying objects are paper and metal. Plastics were still special, not yet synonymous with disposable. In the mid-1950s, the average home might have a rotary telephone and set of serving dishes made of molded Bakelite. The brief article claims, "The objects flying through the air in this picture would take 40 hours to clean—except that no housewife need bother. They are all meant to be thrown away after use." The claim is shameless hype, but we allowed ourselves to be convinced. Mom will be empowered by the trash can, freed from the kitchen sink, and what's good for Mom is good for the whole family. It's women's lib, 1950s style.

We need not mourn the promised disposables that failed to catch on: throwaway draperies, hot pads, feeding dishes for dogs, and disposable goose and duck decoys for the hunting set. Pandora's plastic box was yet to release massive plumes of throwaway pens, lighters, shavers, syringes, gloves, tampon applicators, catheters, and shopping bags; or the bottles, tubs, jugs, trays, tubes, and blister packs containing beverages, foods, pills and products yet to be conceived; or wrappers, pouches, packing peanuts and bubble wrap; or the endless flow of resinous effluvia from fast-food joints and mini-marts. And that doesn't count plasticky items like computers, dumped at a rate of 180 million units each year but technically durable goods meant to last at least three years.

Now when something breaks we throw it away. When was the last time anyone had a broken toaster fixed? It's cheaper to buy a new one. Apple shareholders rejoiced in 2006 when Steve

Jobs urged his minions to buy a new iPod every year, a gadget costing hundreds. The old ones still worked. Now, 300 million plastic-packed iPods (and a few other products) later, Jobs's business success is epic. I find it hard to cheer. In electronic gadgets like these—containing myriad toxic metals as well as waning resources like copper and oil—innovation and disposability join hands for one reason: profit.

LET'S HAVE ANOTHER LOOK at mid-century America, at our way of life around the time *Life* unveiled throwaway living. It was still the postwar era. People were moving from cities and country into suburbs created by a massive surge in home building. The baby boom was on. In the 1950s, Dad could earn enough to support a family. The standard thirst quenchers were milk, juice, and water served in glass tumblers. Mom stirred up paper packets of Kool-Aid for summertime refreshment. On special occasions, Coke, 7-Up, Orange Crush, and ginger ale were poured from deposit-return bottles. Personal products—for example, Lustre-Creme, "the favorite shampoo of four out of five movie stars," and Brylcreem, "A little dab'll do ya"—came in glass jars and bottles or aluminum tubes. Yet to be unleashed was hair conditioner, that ingenious polymer-packed enabler of the daily shampoo that swirls down shower drains and now threatens to coat lake and ocean habitats with surfactant slicks. Indeed, the typical mid-fifties lady paid a weekly visit to the beauty parlor in lieu of attempting home hair care. Fast-food outlets and gas station stop & shops had just begun to pop up here and there. In 1950, the average home measured one thousand square feet and sheltered 3.37 people whose clothes and shoes fit comfortably into closets that, by today's standards, would barely accommodate the underwear. Today the average home boasts more than twice the space for one less person. As nature abhors a vacuum, the space fills with stuff, mostly plastic, that overflows into dumpsters, charity thrift shops, and rental storage units—an industry unknown in the fifties.

Where did all the stuff come from? If necessity is the mother of invention, then war is the most fertile of mothers. Commandeered by the military in 1941, American industry retooled as a war machine the likes of which the world had never seen. The war ended in 1945. Government contracts abruptly dried up, leaving factories spring-loaded for production in a withered domestic market. The American public had some adjusting to do. The Great Depression had forced frugality. World War II demanded patriotic sacrifice. Posters and newsreels touted the virtues of maintaining a "Victory" home by growing and sharing your own vegetables, eating less meat and butter, mending rather than buying new clothes and shoes, and giving all but your most essential metal objects to community drives, to be recycled as planes, tanks, jeeps, and guns for the war effort.

But this story has never been only about plastics. It's about an epic shift from austerity and frugality to abundance and profligacy. When struggle was still the norm, inspirational texts—from the New Testament to Ben Franklin's *Poor Richard's Almanac*—preached thrift and simplicity, thereby ennobling the reality of most people's lives. But writing of his travels in America in the 1830s, Alexis de Tocqueville had reason to observe: "As one digs deeper into the national character of the Americans, one sees that they have sought the value of everything in this world only in the answer to this single question: how much money will it bring in?" De Tocqueville attributed this trait to America's fluid social boundaries and promise of worldly success. The Industrial Revolution spawned the middle class, but true prosperity remained the province of those who controlled resources and production until the late nineteenth century. By century's turn, the economist Thorstein Veblen had invented the term *conspicuous consumption* and applied it to belle-époque nouveau riche, whose purchases were meant to convey status.

In post-Depression, postwar America, the ethical splinter of "Waste not, want not" required excision as the parade of throwaway items marched into supermarkets. Madison Avenue was hardly new to the task. In 1924, Kimberly Clark hired pioneering adman Albert Lasker to promote one of the first truly disposable products: the Kotex pad. Lasker, who famously said: "The products that I like to advertise most are those that are only used once!" The first of modern-day disposables might have been paper collars and cuffs for men, which became commonplace as paper got cheaper during the Civil War era. They were a boon to overworked women, who had plenty to do aside from washing, bleaching, starching, and pressing men's detachable cuffs and collars.

If convenience mothered paper collars, hygiene proved the driver behind toilet paper. Paper towels and cups were beginning to appear in public restrooms by the end of the nineteenth century. Marketing and science proceeded hand in hand with development of the disposable paper cup, introduced in 1909 to replace shared glassware or dippers in schools and other public places. A 1907 study had shown such sharing entailed sharing of germs as well. The dominant product was the Health Kup, rechristened the Dixie cup in 1919. Technology and disposability were similarly entwined. Equipment that could mass produce paper cups, cartons, and bags, and glass bottles and metal cans, arrived in the decades around 1900. As we've seen, several early plastics—notably Bakelite and celluloid—were used to make durable goods like radios, telephones, and film stock before World War II. An economist in the 1920s presciently wrote: "The truth is that while there are definite limits to the possibility of any natural product there are, theoretically at least, no limits to the possibilities of a chemically produced product."

Consumerism was already a hot topic in post–World War I America. In 1933, Herbert Hoover commissioned a report from economist Robert S. Lynd that would help explain the Crash of 1929. It was titled "Things Affecting What People Consume." Lynd wrote that before the crash of '29, marketing professionals had developed "an effective fine art" that enabled them to exploit human frailty to boost consumption. He said marketers viewed "job insecurity, social insecurity, monotony, loneliness, failure to marry, and other situations of tension, as opportunities . . . to elevate more and more commodities to the class of personality buffers. At each exposed point the alert merchandiser is ready with a panacea." He was describing retail therapy.

The "father" of modern marketing was Edward Bernays, whose long career, and life, spanned the twentieth century. A nephew of Sigmund Freud and student of Ivan Pavlov, he drew on both to create a potent arsenal of weapons of mass persuasion still used today. In his words, the goal of marketing is to achieve the "engineering of consent"—that is, to create wants, and to convert wants to needs. But it was all for a good cause. Given a supply glut following World War I, consumer spending was seen as the way to economic stability and prosperity. And it worked . . . until October 25, 1929.

A turning point occurred during the Depression. As a way out of the economic abyss, a widely circulated treatise urged increased production of products designed to need constant replacing. In other words, disposable products. By mid-century, Vance

Packard, the social critic, blamed sophisticated advertising, planned obsolescence, and our ongoing love affairs with novelty and social climbing for rampant consumerism, now defined as buying things in excess of what one really needs. And now, moreover, seen as imperiling America's soul while assuring its prosperity. Writing in 1964, Herbert Marcuse noted a fait accompli when he observed: "The people recognize themselves in their commodities; they find their soul in their automobile, hi-fi set, split-level home, kitchen equipment." We could buy social status. We could spend our way to prosperity. We could even fashion identities by drawing from the "empire of things."

Of course, the biggest consumer item to change our culture was the auto. After World War II, the fifty U.S. factories producing synthetic rubber tires easily retooled for the domestic market. Indeed, working in league with GM and the oil companies, the tire industry helped derail existing mass transit systems in all but a few American cites, safeguarding a market that now consumes one billion tires per year. And following this, as we've seen, the first wave of postwar plastic products featured hard plastic items meant to last. Things like Hula Hoops, Formica countertops, and vinyl records.

Then came the U.S. interstate highway system. It was the largest public works project ever undertaken and the special baby of President Dwight D. Eisenhower. As commander of Allied forces in World War II, Eisenhower had admired Germany's autobahn. Not the least of its virtues was the ease of travel it afforded military vehicles and equipment. When Ike assumed the presidency in 1953, building an interstate highway system was high on his to-do list. In 1956, Congress, at the president's urging and with a lobbying assist from the auto industry, authorized the Federal-Aid Highway Act. Not only would the new roadways enhance the country's military preparedness, the economy would boom as never before.

Well before the Great Depression, American industry had begun to realize benefits from centralized mass production, packaged goods, branding, and advertising. By the twenties, Heinz and Campbell's were producing canned soups and sauces; Quaker and Pillsbury, bagged oats and flour; Colgate and Procter & Gamble, toothpaste and soap. Catalog ordering of clothes and other goods was popular. The centuries-old model of local and home-based production of basic necessities had given way to something that looked more like our way of life today.

But the new interstate highway system literally paved the way for further consolidation of what had been local industries, dairy being an example. It also spelled decline for the rail system and a boom for trucking. In the 1940s, 2,300 farmers' cooperatives provided milk to local markets. By 2002, only 196 remained, with a mere five accounting for nearly half of all dairy production in the United States. (Urbanization, enhanced bovine genetics, improved processing, and refrigerated trucking were factors as well.) The website for an old-time Connecticut dairy, Wades, elaborates: "When supermarkets took the distribution of milk away from home delivery, they forced the use of one-way containers by packaging their private label milk in cartons and by refusing to accept milk from other companies in refillable bottles." Returnable, reusable bottles stopped making financial sense. Local bakeries underwent a similar shift. The Helms bakery truck that visited our neighborhood each week, enticing crowds with the intoxicating scents of fresh-baked bread and warm doughnuts, organized in wooden drawers, with your choice handed to you in waxed paper and paper bags, was gone by the mid-1960s. By then chain supermarkets offered cheap, white

Wonder bread ("Helps build strong bodies 12 ways"), trucked in from the Midwest and kept fresh in its plastic wrapper printed with balloons.

Mass production and distribution of food was trickier than that of other products. Food production could only begin to be centralized when the challenge of perishability was overcome. Glass as a preserving container for food and drink achieved wide use after the first automatic rotary bottle-making machine was patented in 1889. Until the late 1960s, most liquid products, food and non-food, were packaged in glass bottles. Glass remains the preferred material for premium foods like jams, condiments, and better beverages.

The tin can was invented in France, after Napoleon offered twelve thousand francs to anyone who could find a way to preserve food for his troops. Cardboard boxes were introduced by Kellogg's in 1906, conveying its novel new product, cornflakes. A waxed paper liner sealed in freshness. Aluminum foil packaging became common in the 1950s, and aluminum cans found their way to supermarket shelves in 1960.

And then came plastic. The earliest plastic commercial product bottle on record is the PVC squeezable bottle designed in 1947 by Dr. Jules Montenier, inventor of Stopette deodorant. It was the first deodorant spray and proud sponsor of the popular game show *What's My Line?* Montenier's patent application called his invention a "unitary container and atomizer for liquids." Developed with Chicago's Plax Corporation—later to be acquired by Monsanto and then spun off—it showed that mass production blow molding was feasible. Spray-on body powder and Finesse shampoo soon sprang from the fertile mind of Dr. Montenier, all packaged in PVC plastic and promoted on TV.

But plastic film was the material that was to envelop the world by providing a cheap, lightweight, impermeable material in which perishables could be economically shipped over long distances and retain their freshness. Plastic packaging helped liberate food and beverages from local production. And it would usher in the era of imperishable trash.

The first plastic film had been serendipitously discovered at Dow—test tube residue—and soon found use as a moisture barrier for military equipment. But this early incarnation was green and smelly, belying its chemical origins. Refinement was needed before it could earn approval for food packaging in 1953. These days polyethylene film is the single most abundant plastic form. An estimated 80 million metric tons of polyethylene are produced annually worldwide. Most of it is used in packaging. Packaging is by far the greediest user of plastics, consuming a third of total resin production, with, as we've seen, consumer items, institutional products, and construction distant runners-up.

When did the words *plastic* and *disposable* become nearly synonymous? We've seen that plastics were once special enough to be reserved for consumer goods intended for long-term use. Then along came Bic, the French company that acquired rights to the first commercially produced, if imperfect, ballpoint pen, the Biro, made in Argentina. After figuring out how to keep the ink from blobbing, Bic designers opted for a clear polystyrene cylinder with a pinhole to equalize pressure. Having conquered the European and several outlying markets, Bic bought Waterman pens in the United States in 1958. The Bic stick was cheaper than Papermate's ballpoint and wrote as well. It remains second to Papermate in the United States, but is number one in most of the rest of the world. In 2005, Bic produced its 100 billionth Bic Cristal pen. Each day 14 million are sold in 160 countries throughout the world. I don't catch them in my trawls, because hard polystyrene generally sinks. The seabed is likely home to millions.

I do catch felt pens, which float very well, as a hollow tube will. Like a disposable lighter, in fact. The Bic lighter was launched in 1973, the second on the market after Gillette's Cricket. But Bic's lighter was cheaper by half and by 1984 Gillette had given up. Now Bic's main competition is cheap Chinese knockoffs that wholesale for less than a quarter each. Bic sells 250 million lighters per year in the United States and claims title to the top-selling disposable lighter brand in the world. These lighters—three thousand flames guaranteed—and their clones are a lethal staple in the diet of Laysan albatross. A Bic spokesman says the company is "dismayed" by findings of disposable lighters in Laysan albatross chick guts on remote islands, but insists few if any of those lighters were made by Bic. We may eventually find out. A Japanese researcher has put out a call for disposable plastic lighters recovered on beaches around the world. His intention is to trace their origins via markings found on their bodies. The lighters are put to heavy use by the fishing fleet, but the volume recovered on remote beaches suggests terrestrial origins as well. Each Bic product, including disposable shavers, contains between 5 and 6 grams of plastic—not much, only two or three hundred plastic resin pellets to melt and mold. But it adds up when you're producing billions of units each year. Five billion lighters, pens and shavers will generate 30 million tons of plastic that will last longer than anyone reading this—and their children. The market doesn't want your products to be keepers. There is a difference between a product and a possession. A product is ephemeral, something to be used until it runs out. A possession is something to keep, use, and value. Many things that were once possessions are now products. Pens, razors, zippo lighters, and Montblanc pens were once desired and meaningful gifts. No more.

The concept of "planned obsolescence" is crucial to throwaway living. Henry Ford naïvely engineered the Model T to last, thinking the automobile market was unlimited and forever his. Then came General Motors with an entirely different philosophy: new models every year, as well as entirely new lines of cars, playing to the American love of novelty and flash. By the fifties, a darker side of choice had emerged: intentionally, the cars were *not* engineered to last, and had fins added or subtracted and bodies raised or lowered to provoke sales. Other product categories followed suit. Lightbulbs, batteries, iPods—the list is endless— made to last only so long before breaking.

With disposability comes litter, litter being errant eyesore garbage. The plastics industry has a mantra. It is that plastic litter is a "people problem." The tactic has succeeded in de-linking plastics pollution from the material itself and the industries making it. Documentarian and author Heather Rogers tells a revealing story about the early days of disposable containers. In 1953, farmers with seats in the Vermont state legislature pushed through a bill banning non-returnable glass beverage bottles. Motorists were tossing the bottles into roadside fields, where they became a lethal ingredient in cattle fodder. Within months, the major can and glass bottle producers had formed a nonprofit organization and enlisted support from the likes of Coca-Cola, Dixie, and the National Association of Manufacturers. Its name: Keep America Beautiful (KAB).

With deep pockets and media savvy, KAB launched a campaign that reached every nook and cranny of America: "Don't be a litterbug." The litterbug was a species first seen in 1947, when the jitterbug craze was in full swing, appearing on "etiquette" posters around the New York City subway system. Unlike Smokey the Bear, the litterbug had no set image, but always inspired contempt. The covert point of the campaign was to change the subject and to shift blame. To wit: The new throwaway containers weren't the real problem. The real problem was shameful people who didn't responsibly dispose of them. Littering had become a transgressive act, worse, at the time, than chain-smoking in a movie theater and way worse than producing millions of bottles a year that had nowhere to go but a roadside or the dump. The Vermont ban on disposable bottles fell in 1957. KAB's later efforts included the Iron Eyes Cody campaign, featuring the teary Native American who turned out to be the son of Sicilian immigrants (but who lived as a Native American and was accepted as such), and a 1963 "educational" film narrated by none other than Ronald Reagan. He who said of redwood trees, "Seen one, seen 'em all," intoned: "[Trash] becomes litter only after people thoughtlessly discard it."

KAB's efforts may be the first example of "greenwashing." The polluting industries co-opted the role of environmental guardian, defined the new normal, and, yes, engineered consent. Inevitably, by 1965, a nationwide garbage crisis was brewing, leading to passage of the federal Solid Waste Disposal Act of 1965. Municipalities were required to install sanitary landfills (invented in 1937 by the commissioner of public works in Fresno, California) if they didn't already have them. Within a decade it was clear that a new feature of the waste stream—toxic "hazardous" substances—required further action. Congress passed the Resource Conservation and Recovery Act in 1976. The generators of the garbage flow—manufacturers—were off the hook. The consumer/taxpayer—you and me—would pick up the tab in the form of merchandise markup and hauling service or tax-subsidized garbage collection. The state of California alone shells out about $750 million per year of taxpayer funds to landfill just plastics.

Not surprisingly, we at Algalita have had our own brushes with "the industry"—some good, some bad, some amusing. Our grant-funded work with California's State Water Resources Control Board entailed setting up an advisory board and a website. The advisory board included stakeholders from industry, the state, and the environmental community. Since our work focused on learning the true extent of plastic debris in the environment, we decided on the URL plasticdebris.org for the website— despite the pleas of industry reps not to link the words *plastic* and *debris*. They felt it was unfair for plastics to be singled out, since debris can include other materials. Technically, yes, but our findings in both the gyre and near-shore waters showed that plastics account for more than 90 percent of non-natural debris.

With approval from the Water Resources Control Board, we began the registration process for plasticdebris.org, only to learn, to our great surprise, that our intended URL was already taken. Our webmaster made inquiries and reported back that the URL was registered to a person with the last name of Krebs. Well! This happened to be the name of a top executive at the American Plastics Council, the lobbying group for the plastics industry that was soon to be absorbed into the American Chemistry Council. An industry ally on our advisory board managed to talk the American Plastics Council down, and the website is still up and attracting a steady stream of Internet traffic.

This is the first time in history that the material that defines the age is not reused. You can recycle the steel in a car to make a new car. You can sterilize or recycle a glass milk bottle and have a "new" milk bottle, but not a plastic milk bottle, because it's against the law. The melt temperature for PE is too low to ensure proper sterilization. It's not merely disposable: it must be disposed of, or down-cycled to a lesser product like plastic lumber, which is not exactly standing the test of time. Plastic lumber used for picnic tables and a boardwalk-style path at a park in Hawaii was no match for the intense subtropical sun. Within days, heat loading in the material caused the walkway to buckle, and it was replaced by cement.

What if we were to consider just one more thing as those tens of thousands of innovative new products—food, drink, gadgets—are dangled before us each year? What if we were also to consider how many of them, and how much of their packaging, was going to end up in the oceans? I have dedicated much of my life to answering that question. Definite answers are elusive, but I feel confident in saying it's a lot. What if a preconceived endgame for each thing we made or bought was the rule, not the exception? If we had such a plan, then plastics as a material, valued so little and disposed of so readily, would be rehabilitated. The additive cost to the planet of making and disposing of plastics is so overwhelming that nobody wants to think about it. So in our minds we devalue plastics even more—the throwaway products and the wrapping and the bags and the bottles and the tubs—and this relieves our anxiety. Without an endgame, it all becomes mere waste, and waste is worthless.

And with that thought, we return to my adventures as the citizen-scientist, back in port with jars of this worthless waste, now in little bits, culled from the hapless surface waters of the ocean and waiting to be analyzed.

CHAPTER 7

HARM

WE SIEVED NEARLY ONE HUNDRED MILES of ocean surface in swaths three feet wide and six inches deep. We hauled literally a ton of smelly and befouled junk plastic on board, quantified it, and sailed it back to port like spoils of war. We'd like to think we've done the ocean a favor. But what matters most are the samples waiting patiently for quantitative analysis. They're in standard glass quart jars and a few Costco jumbo nut jars large enough to hold several "macro"-plastics, like a soda bottle swallowed by the trawl net. Sealed tight, each holds just enough isopropyl alcohol and seawater to cover the contents. It's pelagic gumbo, an oozy brew of plastic scraps and planktonic tissue. These fragments were hiding in plain sight. You'd never see them from a big ship's deck. Given a slight jiggle, the jars resemble glass snow globes swirling with multicolored plastic snow.

Back in port, Mike Baker and I artfully arrange big gyre debris on *Alguita*'s stern deck for local media photo ops. The reporters are incredulous and aghast. One headline reads: "Plastic Is Drastic" in the mid-Pacific. Afterward, we haul the drastic plastic across the street to my house, where we heap it on a tarp in a shady corner of the backyard patio. In a few days I'll start my Caddy trip to Seattle to meet with Ebbesmeyer. The jars will go to Chuck Mitchell's lab in Costa Mesa, he being the loaner of the superfine manta net. His company, MBC Applied Environmental Sciences, also boasts a full-service marine health diagnostic clinic, and he has generously offered use of essential analytical equipment: dissecting microscopes and electronic scales.

We kept to the collection plan developed by Molly Leecaster and Curtis Ebbesmeyer, but we wound up with eleven samples, not twelve. What? To think we knocked off for a day after five days of sampling and got home two days early—but one sample short! All we can do is chalk it up to another "mystery of the gyre." Leecaster assures us this deviation won't compromise the integrity of the study.

When I return from my quick swing up to Seattle, it's still September. My mother is glad to see me back in one piece and the Caddy in better shape than ever. Those of us who planned, executed, and supported the expedition share a sense that the contents of these eleven jars will shake the world. Steve Ignell's "We already did that" still echoes, attesting to the reason I founded Algalita: to shorten the distance between discovery and remediation of marine contamination, no matter the contaminant. They already did that, but nothing happened. Nothing changed because no one got upset enough. Findings wind up in science journals and conference programs, but ensuing calls to action, when they occur, are strangely ineffectual, perhaps because they're mostly unheard.

Unfettered and independent, our modest but focused little

foundation embraces DIY (do-it-yourself) science and supports others who advance our mission, which at this juncture happens to be eco-diagnostics. The way I see it, our work hews to an entrepreneurial rather than an agency or academic model. We're free of influence from corporate funders with commercial agendas and we're immune to institutional apathy. We have no profit motive, only a philosophical commitment to protect the oceans from the indestructible plastic slag generated by supercharged scientific, technical, and economic "progress."

But because Algalita has an agenda, we're acutely aware that our science has to be rock solid. With diagnostics accomplished, why and how this mid-ocean plastics cemetery came to be will perhaps become more comprehensible. And then we intend to act as driver of the cure. If sources of the plastics can be identified, remedies should reveal themselves. But we're not naïve. I find the challenge motivating. If our results are what we think they'll be, we'll make noise and be heard. Good people will be moved to action, and eventually—when, no one can say—the ocean's plastic burden will be lightened. All predicated, of course, on the science.

Steve Weisberg at SCCWRP makes time for a quick debriefing. I show him one of the jars, plastic soup sloshing inside. "Whoa," he says. "This could be explosive." We're thinking the same thing: that these are significant findings; that the analysis will show that plastics have managed to pollute even the remote mid-ocean; that the results will likely merit publication in a peer-reviewed scientific journal. I'm plenty disturbed by what we saw and gathered out there, because it's ugly and it's wrong. But I'm also willing to admit that my outrage thermostat is set a good deal lower than the average person's. Not many seafarers have seen both the pristine "before" and the befouled "after" of the northeastern Pacific. Fewer still have nerves raw as mine when it comes to man's wasteful and polluting ways. The aesthetic injury to the oceans is bad enough, but the glaring symbolism of moral failure is equally strong.

SCCWRP has just suffered a funding cut—painful for them, as staff will need to be culled. But it's an opportunity for the Algalita Marine Research Foundation. We grab Ann Zellers, a skillful lab tech whose position has fallen under the ax. Now Algalita has its first staff research biologist. Starting with this study, and continuing to this day, she has processed millions of plastic bits and would deserve a place in the Guinness book of world records if such a category existed. I'm no stranger myself to the doing of science. I grew up around science and know how it works. As I've mentioned, my father was an industrial chemist and head of the Hancock Chemical Company. For fun, he devised a way of making Hawaiian tikis from sulfur, a by-product of petroleum refining. I was a chem major and know my way around a lab. No passion allowed, and no overstepping of bounds . . . even when common sense tempts one to make reasonable assumptions based on strong evidence.

At Chuck Mitchell's state-of-the-art lab, Zellers and I set to work. We aim not only to determine the amount of microplastics in the gyre but to develop a profile of the neuston layer's overall contents. As far as we can tell, it's never been done. During the research voyage, we'd removed the eleven samples from their highly toxic formalin bath and rinsed them in freshwater, then rebottled them with less toxic 70 percent isopropyl (rubbing) alcohol. Now we split each sample to facilitate processing, then put each divided portion into petri dishes containing pure seawater. There they await sorting under a dissecting microscope.

Most of the plastics float, while heavier planktonic tissue sinks, making the job a little easier.

Using stainless-steel tweezers and small spoons, we gingerly remove the plastic top layers and heavier fragments from the bottom, leaving the biological matter. Each fraction is examined under the microscope, which reveals slight cross-contamination. So we pluck microscopic bits of plankton tissue from plastic and microplastic bits from plankton and plankton from extraneous, mostly biological matter (feathers, squid eyes, fish eggs, algae, and some small bits of tar) swept into the trawls. These are labeled with their sample numbers and set aside. We serially pop plastic groups and plankton groups into a special-purpose oven. They dry at around 150 degrees Fahrenheit for twenty-four hours. I question Chuck Mitchell about this protocol for processing biologic tissue. The plastic won't undergo much if any alteration from drying, but bio-tissue will obviously lose mass and weight. Mitchell assures me this is standard procedure, that what's left after baking is considered biomass, the organisms' food value. But he says we can also blot the wet plankton on paper towels, then weigh it to get "blotted wet weight." Comparing blotted wet weight with dry weight tells us how much nonnutritive fluid the plankton holds.

We mindfully borrow Day, Shaw, and Ignell's protocol to sort and classify the plastics. Non-standard protocol is often a bone of contention in disputed science. This means buying six graduated Tyler screens, normally used to size geological samples. First, the larger plastic pieces—including our soda bottle— are removed, weighed, and measured. We rinse the smaller pieces through the screens. Of this group, the largest are the size of checkers, the smallest like grains of sand. The subdividing continues. Within size groups, the plastics are sorted first by category, then color, just as Ignell and his colleagues did before. We amplify our plastic categories to allow for finer distinctions and in hopes of discerning the origins of the fragments. We have miscellaneous fragments, Styrofoam and other foam fragments, pellets (aka nurdles), polypropylene or polyethylene line or net fragments, and what turns out to be the biggest identifiable group, thin plastic films.

I hire plankton experts to count and identify the various planktonic organisms captured in the samples, though we wind up not needing this taxonomic data. A disturbing result comes from counting individual plankton. We find in one sample that plastic chips outnumber them. This should mollify those who might take issue with a weight comparison between solid plastics and desiccated plankton.

The process takes months. The results, indicating an improbably high mass of plastic in the central gyre, are more shocking than we could have imagined. Susan Zoske, who is now our extremely proactive executive director, prepares a press release and is poised by the fax machine when I suggest checking first with our procedural guru, Steve Weisberg. We want impact, but we also want maximum credibility. Timing is crucial, and I think we may be on the verge of pulling the trigger prematurely. Weisberg, contacted, emphatically favors waiting for peer-reviewed validation before making public announcements—not because he doesn't trust the results, but because he knows the pitfalls of the process. Right now we have an "unsubstantiated claim." So Zoske holds off and we consider our options. Soon one presents itself. We will test the waters, somewhat literally, within the marine science community, at an upcoming event.

IT'S FEBRUARY 2000. I push through the doors of the new Price Center at U.C. San Diego, the university I left just short of graduating three decades earlier. My comrades and I organized various protests in the precursor to this glass, cement, and steel fortress, back when it was called the Student Center. And before food service was provided by Burger King, Subway, and Panda Express. I'm guessing food fights are now few and far between, but we had some great ones back in the day.

As I'd walked across the eucalyptus-shaded campus, I noted natural feelings of nostalgia but no regret for leaving during the radical sixties when the school was hosting job fairs featuring Dow Chemical, the maker of napalm. Back then I learned that their polystyrene nurdles were an essential ingredient in this flaming horror of an "anti-personnel" weapon. I'm carrying a briefcase containing graphs and an abstract of my findings in the Northeast Subtropical Gyre. I've signed up for a two-day symposium titled Oceanography: The Making of a Science. The event is co-sponsored by the U.S. Navy, the Scripps Institution of Oceanography, and the H. John Heinz III Center for Science, Economics and the Environment. An array of marine science eminences will gather here. With my "explosive" data, I hope to gain science establishment support and perhaps even research collaborators. Several of the presenters seem likely targets. One is Dr. Edward Goldberg, a titan in the field of marine chemistry and an early predictor of the plastic plague in the ocean. In 1994 he'd written an editorial for the *Marine Pollution Bulletin*, titled "Diamonds and Plastics Are Forever?" In it, he alerted marine scientists to the potential for plastics to cover and choke the sea-

bed, thereby inhibiting carbon sequestration. This remains a hot
climate change issue. By now he's elderly and emeritus, but re-
vered by marine conservationists for establishing the International Mussel Watch, a program that assesses shore water toxicity
by sampling local mussel colonies—marine equivalents of canar-
ies in the mine shaft. Another is Richard "Ricky" Grigg, a cham-
pion surfer, Stanford graduate, and Scripps-trained University of
Hawaii professor. His academic specialty is coral reefs, but his
reputation was made in 1965 when, as a grad student, he lived
250 feet underwater for 45 days in *Sealab II*, sharing quarters
with astronaut Scott Carpenter. Both Grigg and Goldberg are
giving talks.

I approach the two separately during breaks and mention
work being done by Algalita, which they have not heard of. I
pull out a few graphs we've prepared for the occasion. This, I
say, is what we found in the middle of the Northeast Pacific
Gyre. Microplastic bits weighing on average six times more than
plankton in our samples. In one sample, plastic bits outnumbering plankton. A total of 27,484 of plastic flakes, chunks, and
pellets sifted from an eighty-mile-by-three-foot-wide strip of
mid-ocean. The statisticians Molly Leecaster and Shelly Moore
have calculated 2.7 pounds of plastic particles per square mile in
the 62,458-square-mile study area—the size of Wisconsin—
where we pulled samples, an average of 334,271 per square kilometer. This works out to 84.3 tons of tiny little plastic bits in this
circumscribed section of ocean—part of the North Pacific Subtropical Gyre. The calculation does *not* include the macro- and
megaplastics, the nets, crates, floats, bleach bottles, shoes, toothbrushes, and all the other stuff we either dragged on board or
sighted and noted in the log.

I tell Grigg and Goldberg that the big items we'd ferried back to port have been dried, weighed, and inventoried. The plastic debris we sighted but could not retrieve we have also accounted for with descriptions and estimated weight. The single largest feature we came upon was a tangled mass of ghost nets. We estimated the weight at a ton, but there was no chance of recovering it. Total macroplastics—both collected and documented at sea—we estimate at more than two tons. Small stuff weighs in at a shade less than three pounds. All dried plankton from the samples comes in at half a pound. Virtually all the plastic bits are fragments of degraded, formerly intact objects. Few are nurdles, the famous preproduction pellets. This means the preponderance of what we recovered—unidentifiable plastic fragments—was calved from lost and dumped objects. In these fragments, we see the future of the fishing nets, crates, bottle caps, sour cream containers, soda bottles, and a million other things now bobbing in the gyre. It's not unreasonable to predict a quantum leap in the fragment population as these objects degrade.

I've compared our samples with those of Day, Shaw, and Ignell. More than ten years after their study and ten years after MARPOL Annex V, mid-ocean plastic trash in the most polluted areas is now three times more abundant than what they recorded. At this rate, it might not take a century for plastic to coat the ocean surface. I say to Grigg and Goldberg that the central gyre could one day be a virtual floating beach of plastic sand. I don't say that we feel as if we're documenting a slow-mo sci-fi disaster, only there's nothing fictional about it.

The reactions of Grigg and Goldberg are ones I haven't yet seen, but am soon to get used to. They are not upset. They are not excited. Goldberg commends me for doing independent sci-

ence and offers to communicate going forward. I give him a copy
of the data but never hear back. He passes away in 2008, well into
his eighties. Grigg's response shocks me, but in retrospect I come
to appreciate its value. He says I need to show harm.

I point out to Grigg how we have calculated the mass of
plankton as well as that of microplastics. It seems obvious that
plankton-eating filter feeders would have a hard time *not* ingesting it. But it needs to be proved, says Grigg. And even if they are
eating it, what harm is it doing? I say a man-made synthetic is
extremely likely to be entering the food chain: how can there be
anything good about that? He says groups like the UN, the main
international body dealing with non-sovereign ocean issues,
don't care if the ocean is full of plastic. They will want hard evidence of harm before considering policy changes.

I should have known better. I think a plastic garbage patch in
the middle of the Pacific screams harm just by being there. Plastic
garbage does not belong in the ocean any more than sharks belong in municipal swimming pools. Plastic is like an invasive species. Once established, it doesn't go away. The oceans, to a point,
can assimilate pollutants, even oil. But oil that's been catalyzed
and converted into a synthetic form, plastic, doesn't dissipate. It
accumulates. Its presence on earth grows by 300 million tons each
year. Even if the fraction that makes its way to the ocean is 5
percent, or 1 percent, or even half a percent, it's still major ton-
nage. Big pieces break down into smaller, more ingestible pieces.
We recovered many objects bearing nibble marks. This insidious
debris represents man's despoilment of the earth's most pristine
environment by one of its least valued materials.

And yet . . . I see that my study of the gyre is not the bomb-
shell that will get the science world's attention. It's a first step.
My work has just begun.

Night trawl sample, myctophids and plastic, Eastern Garbage Patch, 2008. *Jeffery Ernst, AMRF*

RIGHT: Baby sea turtle, on its way out to sea, navigating plastic at Kamilo Beach, 2009.
Janiece Tanner-West

BELOW: Ghost net attracts fish and diver (Captain Charles Moore), North Pacific Central Gyre, 2009.
Lindsey Hoshaw, AMRF

Black-footed albatross on Kure Atoll, 2002.
Cynthia Vanderlip, AMRF

ORV *Alguita* hove to for a swim and a photo.
Lindsey Hoshaw, AMRF

Average sea surface conditions during first gyre sampling, 1999.
James Ingraham Jr., NOAA. Published in the paper "A Comparison of Plastic and Plankton in the North Pacific Central Gyre," in Marine Pollution Bulletin

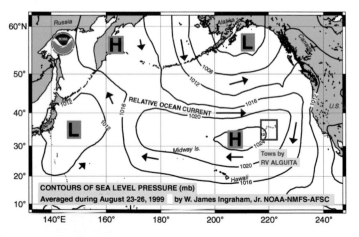

Captain Charles Moore,
Marcus Eriksen, and Jody
Lemmon deploying manta
trawl, 2005 gyre voyage.
Laurie Harvey, AMRF

Curtis Ebbesmeyer
beachcombing at Kamilo,
2007. *Captain Charles Moore*

Gyre sample, 2002.
Matt Cramer, AMRF

Collected gyre
debris with
barnacles, 2005.
Jody Lemmon, AMRF

Chemical industry bottle
cap from the gyre, 2009.
Jeffery Ernst, AMRF

Kamilo Beach plastic sand,
2007. *Jeffery Ernst, AMRF*

Contents of dead
Laysan albatross
chick's stomach,
mostly bottle caps,
Kure Atoll, 2002.
Cynthia Vanderlip,
AMRF

RIGHT: Alan Walti with spilled nurdles at
plastic processor, 2004. *Captain Charles Moore*

BELOW: Laysan albatross, Kure Atoll,
2002. *Cynthia Vanderlip, AMRF*

Contents of dead Laysan albatross chick's stomach, mostly bottle caps, Kure Atoll, 2002.
Cynthia Vanderlip, AMRF

Captain Moore's Kamilo collection, 2007.
Captain Charles Moore

Masked boobies and flotsam on Kure Atoll, 2002.
Cynthia Vanderlip, AMRF

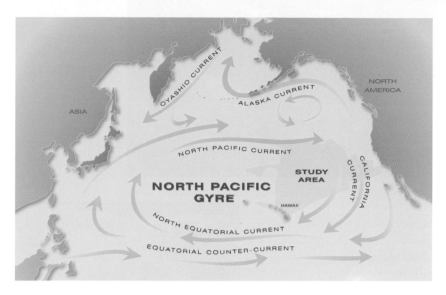

Gyre diagram showing *Alguita* study area. *Jean Kent Unatin, AMRF*

Manta trawl night catch, Eastern Garbage Patch, 2008. *Captain Charles Moore*

Contents of single myctophid stomach, eighty-three plastic fragments, compared with natural food, 2008. *Jeffery Ernst, AMRF*

Captain Moore cutting off ghost net from fouled propeller, 2009.
Scuba Drew Wheeler, AMRF

RIGHT: Sorted bottle caps from Coastal Cleanup Day, Long Beach Peninsula, 2005.
Captain Charles Moore

BELOW: Monk seal in ghost net on Kure Atoll, 2002.
Cynthia Vanderlip, AMRF

I RETURN TO LONG BEACH slightly dispirited but no less resolved. After all, I've only just begun drafting the scientific paper we will submit to the *Marine Pollution Bulletin*, a venerable yet lively scientific journal based in England. Published peer-reviewed studies are the gold standard of scientific credibility, and it's rare that anyone other than Ph.D.s, government researchers, or graduate students see their names in science journals. I have asked Steve Weisberg as well as the statisticians Molly Leecaster and Shelly Moore of the SCCWRP to stand by me as coauthors. I will write up the results, but their editorial guidance will be essential. Not only did they offer material help with the study plan, but their positions and degrees will enhance the paper's standing. But first I must tell them what Ricky Grigg said and ask them how we go about showing harm.

I bring my first draft to a meeting with Steve Weisberg and tell him about my experience at the symposium. He's unsurprised by the reaction. In retrospect, I know he's also bemused by my naïveté. I show him a rough outline of the paper and he points me toward the SCCWRP library with an assignment: read the journals and learn the article format. It's not like writing an article for the Algalita newsletter. You need to adhere to a strict formula: abstract, introduction, methodology, results, discussion, conclusion. You need to cite other studies germane to your research and list them at the end. You have to be very careful not to overstate what your research has found.

But before sending me on my way, Weisberg has an idea that addresses the issue of harm. My objective has been to provide baselines for plastic and non-plastic matter in the North Pacific Central Gyre. Weisberg says the data are impressive but thinks it's not enough just to show you found X amount of plastic per square mile out there — even if the results show plastic mass *growing* in the decade since MARPOL Annex V. People could say, well, that's a shame, and never think about it again, because problems don't get much more out of sight, out of mind, than this. If we want to make it "pop," he says, we need "context." Weisberg calls me a "tweener," meaning I'm doing legitimate science, but I'm an "outsider" and I've got an agenda. That's fair enough. So he says: Why not just go for it? You've got the plastic count and the plankton count. Compare the two and frame the study as showing the potential for ingestion by filter feeders, the little organisms just above plankton on the food chain, the ones that eat plankton.

It didn't seem like such a big deal at the time, and I'm almost surprised in retrospect that this direct comparison hadn't already occurred to us. As I listen to Weisberg, the challenge is clarified. It can't only be about the ugliness or wrongness of plastic garbage in the remote ocean. And it can't only be about getting baseline measurements to prove and measure its presence. It's about giving credence to the sense that plastic is *doing* something out there, something very possibly unhealthy, something *harmful*. Weisberg's suggestion will set our study apart from earlier assessments and stir up a bigger impact. You're not going to find this much plastic in the middle of nowhere on land — if, say, you vacuumed a three-foot-by-eighty-mile stretch of open desert in Utah. Close to a highway or populated area you might, but not in a wilderness. But in the middle of the north-central Pacific, there it is, in a place — a marine wilderness — on a far greater order of remoteness than a Utah desert. A thousand miles from the closest population center, a distance from humanity not even possible to achieve on land. If you sifted Utah desert equivalent

to all eleven gyre trawls, I doubt you'd find as much plastic content as we captured in one gyre sample. So we think we have a
way, with this plastic-plankton comparison, to put an edge on
the data. And we already have the analysis showing that the
plastic-to-zooplankton ratio is an astounding 6 to 1.

Despite its reputation as an oligotrophic area—poor in nutrients, skimpy on marine life—it's not as if the Central Gyre is a
wasteland. Our trawls caught plenty of zooplankton, including
those tubular filter feeders, the salps, that pump their way through
the water column, randomly feeding on whatever small things lie
in their path, be they plankton or plastic. These creatures populate the lower rungs of the marine food chain, putting them near
ground zero for all things edible on the planet. Just looking at
the sample in the bottle, containing both plankton and plastic,
it's clear as day—if not yet "proven"—that the problem with
plastics goes beyond the perils of seabird ingestion and accidental entanglement, which I'm soon to learn are already well studied. If surface waters of the ocean are a sort of buffet, then the
main course, even for the smallest creatures, could well be plastic
bits that are as much a part of their lives as plastic stuff is in ours.
Then they get eaten. And what eats them is eaten, and so on.

We will look at plastic as if it were "biomass." This will link
the food web with man's plastic garbage in a scientifically defensible way. Claims of ingestion can't be made, as ingestion was
beyond the scope of our study, but the *potential* for ingestion can
be noted. We will mention the large salp we caught in one of the
samples, a *Thetys vagina*. Yes, really. It's that hand-sized sticky
tube of clear jelly studded inside and out with plastic fragments
and bits of polypropylene fishing line, like a sad little float from
a polluted parade.

SO HARM, or at least the potential for harm, sets the agenda for writing the paper. I feel a certain regressive anxiety about producing work that's up to snuff, but I'm happy to be heading as an invited guest to the SCCWRP library. It will be my de facto home for a few days. I'll figure out how to write this paper, and I'll glean what I can about harm being done by plastics pollution. Again, given how little I'd heard about the marine debris issue — despite my years of environmental work — it's more than a little surprising to see how much there is.

The studies begin with Laysan albatross in the 1960s. They include the northern fur seal, marine turtles, an array of seabirds aside from the albatross, and a host of marine mammals. Nineteen eighty-seven is a big year for wildlife impact studies. Perhaps it's no accident that MARPOL Annex V, prohibiting ocean dumping of plastics and other garbage in oceans and oceanbound waterways, is implemented the following year. In 1997, David Laist, a staff biologist with the Marine Mammal Commission, compiled a list of 253 marine species, confirmed victims of entanglement in and/or ingestion of plastics. I also review the study recommended before the voyage by Curtis Ebbesmeyer. Published in 1989, it's by Anthony Andrady, the North Carolina–based materials engineer. His paper is the first to delve into the mechanics of plastics breaking down in the marine environment. Using institutional privilege, Shelly Moore arranges for the proceedings of the NOAA's Third International Conference on Marine Debris, held in 1994 in Miami, to be sent to SCCWRP. I'm happy to have access to this relevant, surprising, and often alarming material, and I don't mind spending the better part of a day it takes to photocopy every one of the articles.

What I'm taking from my review of the literature is that science, far from being cut and dried, can be a shimmery affair. "Scientifically proven" results are rarely exempt from debate, in the form of letters to the journal editor, reinterpretation, bias, disproof, protocol disputes, politics, and failures to replicate. And then there's that baffling disconnect between findings and effective remedial action.

The bar for "harm" is high indeed. There's little debate about harm done by derelict nets and fishing line that strangle and drown through "ghost fishing." But the matter of harm via ingestion is less clear-cut. Studies tell us that seabirds that eat plastics are thinner and less successful procreators than counterparts that have managed to avoid the non-digestible material. We know various species have favorite plastic colors—obviously colors resembling natural food. The Laysan albatross appears to favor red disposable lighters and bottle caps, probably because they resemble squid, an albatross favorite.

I do not see in these studies a call for a reduction in disposable lighter production or for these lighters to be made from a digestible material or in colors other than red. Nor are there demands for a "leash law" to keep the caps on plastic bottles, like the pop tops on soda and beer cans. By the late 1990s, upwards of one hundred thousand Laysan albatross chicks are dying each year at their main rookery on Midway Island, northwest of Hawaii. Their corpses rot, invariably exposing plastic objects in their abdominal cavities. The chicks are ephemera. Their contents are forever. But no study says conclusively that the chicks are outright killed by the plastics because other lethal factors cannot be excluded. Lead in old military buildings, for example, or a parent bird failing to show up with food. Unless all other variables can be eliminated, there's little chance for proof of harm. Furthermore, if harm is done to a non-human, non-threatened, or non-endangered species, does it count as harm worth acting on, especially if there's an economic cost?

I do notice, however, that the focus of most studies is on so-called apex predators: seals, seabirds, turtles, and cetaceans (dolphins and whales). Thus, we know the top of the marine food chain is very well acquainted with plastic trash in the oceans. But it appears the bottom of the food chain is virgin territory for study.

This leads to the first lines of the study's abstract: "The potential for ingestion of plastic particles by open ocean filter feeders was assessed by measuring the relative abundance and mass of neustonic plastic and plankton in surface waters." Our first paragraph asserts rather boldly that "marine debris is more than an aesthetic problem, posing a danger to marine organisms through ingestion and entanglement." The results also seem to validate the oceanographic theory that flotsam accumulates in remote mid-ocean whorls that are soon to be known as garbage patches. The biological ramifications of this phenomenon demand further investigation.

I wasn't the first to be disturbed about plastic trash in the ocean, and I wasn't the first to study it. But maybe I was the first to freak out about it. But will I—an uncredentialed outsider, lacking standing in a rigorous field—be heard? I see clearly that Steve Weisberg is making a difference, with science that serves policy makers. He would lose credibility—and his job—were he to get on a soapbox and preach ecosystem collapse if we don't change our polluting ways. I'm not constrained this way, but I lack the authority that comes with a Ph.D. and dozens of publications. I don't even have an undergraduate degree. So this entrepreneurial science will need to achieve three ends. It will need to establish my science bona fides so I will be listened to. It will need to show that small as well as large pieces of plastic garbage are rampant in a once pristine marine environment and are possibly entering the food chain. And it will need to propel a campaign, a crusade, now taking shape in my mind. I intend to raise up mid-ocean plastic garbage from out-of-sight-out-of-mind obscurity to front-burner prominence, or at least give it my best shot. You want harm? I'll show you harm. Not just to the oceans but to the planet, to our bodies, and to our very souls.

CHAPTER 8

THE PLASTIC AGE

And Man created the plastic bag and the tin and aluminum can and the cellophane wrapper and the paper plate, and this was good because Man could then take his automobile and buy all his food in one place and He could save that which was good to eat in the refrigerator and throw away that which had no further use. And soon the earth was covered with plastic bags and aluminum cans and paper plates and disposable bottles and there was nowhere to sit down or walk, and Man shook his head and cried: "Look at this Godawful mess."

—ART BUCHWALD, 1970

IT'S A LATE SUMMER DAY, the day of the Ocean Conservancy's annual International Coastal Cleanup. The Ocean Conservancy is an ambitious nonprofit that's made beach litter a global cause célèbre. As we'll soon see, it keeps statistics that offer a "snapshot" of the sorts of things people leave behind at the beach. Along stretches of coast less frequented by visitors, it tallies things the ocean lobs back on the shore. The cleanups are astutely scheduled in late September, after summer folk have returned to work and school and the shorelines are probably as littered as they get. They also beat winter surf and storms that could very well pull residual summer litter into ocean waters.

We're at a broad white sand beach lined by million-dollar homes. A beach popular with families on weekends and volley-ball players. There's not a lot of visible trash on the sand, though the strand line, as usual, gleams with nurdles and bits of plastic. The thought occurs: just what are all these volunteers going to clean up? I decide to conduct a little experiment and show the younger set that science can be fun (and sometimes profitable). I challenge about half a dozen kids, there for the cleanup, to collect bottle caps only. I hand out used plastic shopping bags and prom-ise a nickel for each cap they find. The total cost will be about $20, I figure, assuming they'll scavenge maybe 400. But these kids are avid, focused, competitive, and hell-bent on making a few bucks. An hour and a half later, their haul has me forking over close to $60. From 300 or so meters of broad, clean-looking beach, these incentivized plastic hounds gathered nearly 1,100 caps, none attached to a bottle. It's like trawling surface waters in the mid-Pacific and getting ten times the plastic bits you thought were out there. The responsible citizens who'd left the beach with their redeemable bottles—California is a bottle bill state—walked away from the worthless caps. Among intact ob-jects I find in the open ocean, polypropylene bottle caps are num-ber one; ditto for recognizable objects found in the stomachs of dead Laysan albatross chicks.

Bottle caps reside in a packaging trade category called caps and closures. Once there weren't that many, mostly metal caps and lids from beverages and foods bottled in glass. But now lots and lots are made each year. If you want to know the real skinny on packaging, there's no substitute for industry trade publications, now accessible on the Internet. In early 2011, a headline on foodproductiondaily.com announced the results of a new market analysis: "Caps and closures market to approach $40 bn by 2014." Many if not most of these will be produced in cheap-labor countries and cost almost nothing. I did a nurdle count outside the Alcoa bottle cap factory in Ensenada, Baja California, and met with the management to try to get a tour, but was turned down. We can reasonably estimate about a trillion new caps and closures being produced per year.

The caps and closures study was conducted by the Freedonia Group, a "leading international business research company" based in Cleveland. Ninety percent of Fortune 500 companies avail themselves of Freedonia research, which is anything but free at mid-four figures per individual study. The essence of this report is that the Asia-Pacific region—China especially—will drive growth, while North America's gain will be modest and mostly in caps and closures for pharmaceuticals and foods—not beverages. From my point of view, it's encouraging. The over-all 4.6 percent growth rate is lower than the 6.3 percent rate between 2003 and 2009. So growth overall is slowing, a good thing. But there's a catch. Demand for *plastic* caps and closures is accelerating as the need for metal caps and closures shrinks. The reason, the report says, is that "plastic packaging is gaining at the expense of glass bottles and jars," which typically have those metal caps and lids.

More plastic caps and closures. More plastic containers and bottles. More plastic in China, where trash management ranges from good to none. A bright note is "deceleration" in the bottled water sector due to "environmental concerns" in the United States and Europe. Closures for the food, pharmaceutical, and supplement markets are less likely to pollute the oceans—though I once found a bottle of Viagra (empty) among trash in the Los Angeles River. But they are rarely recycled, making them an injudicious use of dwindling petroleum resources.

Bottle caps are indicator items. Much of what they seal is portable, single-use beverage containers. In the Beatles era, you might have hallucinated a single-serve bottle of water, but you couldn't have bought one, or any single-serve beverage, in plastic. Now you can choose from among 50 billion bottles of water produced annually. Zero plastic shopping bags in 1970 became 500 billion a year by 2011; some say a trillion.

Plastics are both cause and effect of economic growth. Plastics, lumped with the chemical industry, is a top five U.S. manufacturing industry (despite a 30 percent loss to offshoring and automation since its peak in the early 2000s). Worldwide, the closely related but much larger packaging industry is the third largest, after food and energy. Relative to other mega-industries, it whispers. Companies that make packaging don't advertise their wares to the public, and most conglomerates like H.J. Heinz and Procter & Gamble design packaging in-house and often export its production. The end user is the retailer, not the consumer, who doesn't buy the package but the product within, though the package is hardly irrelevant. Beyond containing the product, the crucial job of the package is to lure the shopper to the product.

With the exception of farmers' market produce, most every product we buy is packaged or containerized or both—as in a jar of face cream or bag of Wheaties packed inside boxes. Plastics account for 53 percent of the material used in packaging, though by weight paper has a substantial edge. In the United States a third of garbage landfilled annually is packaging—83 million tons, equivalent to 6.9 million (empty) semitrucks. It's the largest category of solid waste, according to the EPA. Packaging is a science. Michigan State University has an autonomous School of Packaging, amply funded by consumer goods giants, and the Rochester Institute of Technology conducts NASA-funded research into advanced polymers with which to swaddle equipment and supplies sent into space.

Do we even have to ask why trash generation in the United States rose from a daily 2.68 pounds per person in 1960 to 4.5 pounds in 2008? In 1960, total municipal solid waste (MSW) generated was 88 million tons. In 2008 the figure was 250 million tons, and that's with an EPA-reported recycling rate of about 30 percent. Plastic recycling rates vary wildly among regions, but they average a pitiful 13.2 percent, making plastic by far the least recycled packaging material. Leading the pack are paper and paperboard (65.5 percent), followed by steel and aluminum (over 50 percent) and glass (31.3 percent). In 1960, plastics were less than one half of 1 percent of total waste generated. By 1980 they'd risen to 4.5 percent. In 2008, plastics accounted for 12 percent of the waste stream by weight. The EPA gives no figure for the more telling statistic: volume. But a clue is provided by the state of California, which ranked plastics second in volume among materials landfilled. If efforts to compost green waste continue, plastics will ascend to the top spot.

The Institute of Packaging Professionals says the job of a package is to "contain, protect, preserve, transport, inform, and sell." Other experts note that portion control has become an important facet of packaging. Who pays for trillions of dollars' worth of packaging and its disposal? The consumer/taxpayer. Who profits? Manufacturers, investors, and increasingly privatized handlers of the municipal waste stream. In business lingo, packaging is an "externalized" cost picked up by the consumer. The nation's biggest garbage company, Waste Management, Inc., ranks 196 on the Fortune 500, has assets—including landfills in nearly every state—worth $21 billion, and made a $1 billion profit in 2010. If you haven't noticed, garbage has been commoditized. Companies like Waste Management are crucial to our economic system. If our trash weren't instantly whisked away—if it buried our houses and spilled into the streets—we might change our throwaway ways. But out of sight is out of mind, so the cycle of consumption and disposal is vigorously, and profitably, defended and maintained.

We've seen how World War II created a production apparatus beyond any the world had ever seen, and how its capacity far outstripped postwar, post-Depression consumer demand. It fell to the marketing masterminds of industry to shake Americans out of their frugal torpor and ignite a justifiable spirit of getting and spending. Their greatest stroke of genius was product proliferation. Ever-expanding variants of core products—toothpaste, shampoo, cold cereal, canned soup, whatever—began to crowd supermarket shelves, most of them new and/or improved variations of the boring old product. All presented in bright shiny packages. With TV's rapid rise, marketers had a powerful new tool with which to implant new cravings for novelty; new norms for health, hygiene, and child care; new standards for beauty and grooming; and grist for American social ambitions.

Take the classic brand, Welch's grape juice. By the 1950s, Welch's had added grape jelly to its line, as well as sparkling grape juice in both white and purple. The jelly came in collectible tumblers, starting with Howdy Doody in 1950, and the juice aligned with the Mickey Mouse Club in 1955, the jelly with Pokémon in 2002, and by 2003 the juice came in single-serve polypropylene bot-

tles and "Welchito" pouches. The jellies became "spreads" in
"squeezable" plastic bottles that would boost revenues by 50 percent in 2002. But that's not all. Processed fruit snacks and fresh
fruit—bunches of real grapes bagged in methane-eating plastic
film—were still to come. The "wellness movement" ushered in
organic juices, "superfruit" blends, and juices with additives like
fiber and calcium. The company now sells four hundred products
in thirty-five countries. Welch's is a proud, solid American firm,
created in 1869 when a New Jersey doctor named Thomas Bram-
well Welch found a way to pasteurize and bottle "unfermented
wine" for church use. This product never took off, but Welch's
son found a sound secular market for the juice and, like Kellogg
with his flakes, touted its health-giving benefits. Since 1956,
Welch's has operated as a farmer-owned cooperative, and it has
survived in good years and bad because it proliferated its product
line and understood that kids and health sell.

By the time the 1950s rolled around, disposable products like
the ones in *Life*'s "throwaway living" photo were pitched as convenient, time-saving, and hygienic. It was all upside. Value had
shifted from a penny saved to time saved. Disposable products
kept factories humming, workers working, profits growing, and
families reveling in a renewed sense of abundance and upward
mobility. Images of bread lines and near-empty store shelves in
the USSR were ubiquitous in mid-century U.S. news reports,
connecting abundance with free-market capitalism and depriva-
tion with dreaded communism. Between 1950 and 1960, the U.S.
GNP grew nearly 70 percent, fattened by population growth,
manufacturing, new housing, Korean and cold war defense spending, and the release of "pent-up" consumer demand. Women began
flooding into the workplace in the late 1960s. Time grew scarcer,
convenience more highly valued. More meals were taken on

the run, meals that often came in disposable containers—not only take-out, but meals from the deli and freezer cases that only needed heating before eating.

Fast food is nothing new. Street vendors sold bread and wine in ancient Rome. Medieval pilgrims picked up turnovers and buns en route to the Holy Land. The English have enjoyed take-out fish and chips since the late 1800s, and White Castle started its regional slider empire in 1921. But old-time meals on the run didn't come in foam clamshells with polystyrene utensils, weren't chased by beverages poured into PE-impregnated paper cups, imbibed through polypropylene plastic straws of the type collected by the millions in annual beach cleanups. When convenience foods first hit the shelves, new items like Swanson and Banquet TV dinners and pot pies, introduced in the 1950s, felt exotic. Though the food seemed, at best, an approximation of the real thing, it was still strangely special. My thrifty mother washed and stacked the sectional aluminum plates and pie tins in a cupboard till she finally realized the futility of the effort and chucked them. Other frozen foods began to appear in the home freezer—french fries, fish sticks, and "cardboard" pizzas—and suddenly families had a separate, coffin-sized freezer humming in the garage.

That was when the idea of value was still linked to bulk. You got more for your money buying a big box of cereal rather than the mini-pack assortment. But no more. Research conducted by the Hartman Group, food industry consultants based in Bellevue, Washington, indicates that as never before, we equate time with money. Time spent on food preparation is now considered time wasted. Meals from scratch using fresh food are fast becoming the province of effete foodies. Moreover, we've untethered ourselves from the family dinner table. The Hartman report cites increasing "fragmentation" in our eating habits. In families, individual tastes and schedules are honored, leading to more solitary meals often taken in the company of a TV or computer screen. Home eating has been "restaurantized," with choice reflecting personal preference the expectation. Households are getting smaller, with people living alone now the fastest-growing demographic group. By the 1980s, Stouffer's and ConAgra's Healthy Choice "gourmet" meals could be found in the freezer section, arrayed on microwavable plastic plates. For a family, these meals are budget breakers. But they make sense for living-aloners and dieters opting for portion-controlled prefab meals.

The food industry is forever looking for that golden beast, the value-added item. It's a kind of alchemy. You take cheap ordinary ingredients like grain, sugar, and fat, add a few shreds of fiber, a few fortifying minerals, and wrap it all in a shiny package that proclaims life-changing benefits. When food is engineered into complex forms—say, a "power" bar or chicken cordon bleu—the profit margin can be staggering . . . and the garbage trail as well. You can also trust the food industry to pick up on the latest "scientific" findings—oat bran! calcium! vitamin D! zero trans fats!—and create a new product line, requiring special packaging, of course.

Back in the mid-Pacific, my research turns sociological. I'm one for stocking the ship with as much fresh food as can reasonably be expected to last, and ending the day with a fine supper—happily prepared by me—and the pleasure of the communal table. In recent years, however, I find that younger crew members often opt out, preferring to snack their way through the day: an orange here, a muffin there, and popcorn munched while viewing a DVD in the evening. Lots of popcorn—microwaved until we learned about the toxic grease barrier used in the bags—

now popped on the propane burner. I've realized so-called Gen Y grew up in a different paradigm, one in which both parents worked and family members fended for themselves at mealtime. I see these habits as a new point on the continuum that began in the 1950s. I recall when 7-Elevens started popping up on street corners and suddenly gas stations had mini-marts. It seems strange, but 7-Eleven is the largest franchise in the world, bigger than McDonald's, with more than thirty-six thousand stores in eighteen countries. Its success mirrors the way people live, from snack to snack and on the run.

The U.S. economy is now considered "mature." Growth rates in all food/beverage-related sectors are at this point modest compared to those before 2000. But elsewhere in the world this is not the case. India, for example, has nowhere to go but up. Foreign investment is streaming into the country, much of it to help grow the food-processing industry, which is expected to double in the next five years, according to a report at foodproductiondaily .com. More food processing means more food packaging, mostly plastic. The world population will soon reach seven billion, all of whom must eat. It's a challenge, and it's a peerless opportunity for the food industry.

To walk around a supermarket, a discount warehouse, even a natural-foods store, is to enter a world of polyethylene and its kin. In the produce section, at least half the fruits and vegetables come pre-bagged or wrapped in PE film, some treated to eat ethylene off-gassing and prolong shelf life—a bona fide irony. The rest goes into clear PE bags torn from a roll, a breakthrough product introduced in 1966, more than a decade before plastic handle bags displaced paper at the checkout counter. There's no escaping plastic packaging in the obvious places: the bakery, meats, dairy, the beverage aisle, the pharmacy (where polymers coat many medications, especially time-release formulae), and, of course, among personal and cleaning products. In the freezer sections, where paper packaging seems to gain on plastic, look again. The paper is impregnated with PE to impart moisture resistance. Most canned goods are still lined with the epoxy polymer BPA. You might think boxed breakfast cereal is exempt, but box liners are typically high-density PE or glassine, a thin, highly compressed form of paper typically infused with wax; that is, paraffin; that is, a product that comes from the same petro fount as PE.

The uses for PE film extend far beyond food. In agriculture it covers greenhouses and is shade cloth, weed mat, and black or clear plastic "mulch." In transport it protects equipment and wraps loads on pallets. It's a moisture barrier in construction and it's used to line aboveground pools as well as hundred-acre land-fills. One of the earliest commercial uses of low-density polyethylene film was to protectively cover clothes for the trip home from the dry cleaner. An unintended and tragic consequence was a rash of suffocation deaths in children. When a United States Heath Department survey in 1959 revealed the suffocation deaths of sixty-one children in a sixth-month period, the *Journal of the American Medical Association* committee on toxicology investigated. It found the principle cause of accidental death occurred among infants when dry-cleaning bags were reused as waterproof sheeting in their cribs. An aggressive and effective public health campaign was launched. The Consumer Public Safety Commission now keeps accidental death statistics. Plastic bags still cause on average twenty-five deaths a year, with the typical scenario being an infant rolling off a bed and onto a plastic bag containing clothes. Be warned.

The oceans' pollutants reflect the patterns of plastic use on land. Polyethylene film scraps dominate my samples. More than any other plastic type, polyfilm gets around, for reasons most obvious. Each bit of plastic film, most especially flyweight shopping bags (aka urban tumbleweeds), is like a mini-sail just waiting for the next breath of wind. Much loss occurs within the trash collection system. Lidless and overflowing public garbage cans—and sometimes garbage trucks and landfills themselves—are virtual plastic litter disseminators. As I mentioned earlier, as a boy in the fifties I used to visit dump sites with my father, whose natural curiosity took him to odd places. Back then, the garbage just sat there. Now dump sites are alive with dancing plastics, whipping like jazz hands, plastering themselves to chain-link fences, tumbling over barriers into the surrounding countryside. As we'll see, terrestrial as well as marine animals suffer the consequences.

The state of California came close to banning ultra-light shopping bags in 2010, but strenuous push-back from the American Chemistry Council managed to derail the effort by lobbying state senators. Caltrans, the agency charged with maintaining California's freeways, spends $16 million per year to remove just plastic bags from roadways. California has figures suggesting 19 billion single-use bags are handed out yearly in the state, with only 5 percent recycled. A study in Bangladesh found that 9.3 million bags wafted daily into the streets, where they clogged storm drains, intensified monsoon flooding, and promoted deadly waterborne diseases. The country banned the bag in 2002. Thin bags are banned in Mumbai, and in China, South Africa, Eritrea, Rwanda, Somalia, Tanzania, Kenya, and Uganda.

Even as the tide appears to be turning against the twin icons of plastic waste—flimsy plastic shopping bags and single-serve water bottles—and even if they've been attacked ad nauseam, they must be mentioned by dint of their ongoing profusion. Few products have ever gone from boon to bane as quickly as single-serve bottled water. Ironically, health concerns appear to have spawned its ascent. A 1972 EPA report found safety issues with some municipal tap water. Environmental groups took up the cause, and the message was that bottled water—then mostly in gallon jugs and five-gallon glass or polycarbonate watercooler bottles—was a safer bet. In the early eighties the fitness movement, personified by Jane Fonda, urged cleansing hydration. Eight "glasses" of water a day—a problem when you're on the run, solved by a quart-sized Evian bottle. Beauty magazines jumped on the bandwagon with promises of dewy, eternally young skin conferred by constant water intake. In Europe, France especially, restaurants routinely offered bottled water, with or without "gas." As sophisticated European ways drifted into mainstream America, bottled water would naturally be embraced. Given the minimal cost of water, it didn't take long for beverage companies—and resin suppliers—to jump on the trend.

Growth was exponential. In 1985, Americans drank on average 5.4 gallons of bottled water per year. This would have been largely imports and watercooler water. Only five years later, in 1990, the figure had almost doubled to 9.2 gallons per person. Coke and Pepsi jumped into the pool in the mid-1990s, with Dasani and Aquafina. By 2000 another near doubling had occurred, to 17.8 gallons. Then came a quantum leap between 2000 and 2006, when bottled water consumption hit 27.6 gallons per American per year—in more than a billion plastic bottles. What had seemed such a good idea—clean, pure water—started to seem a bad one when the bottles began turning up in staggering numbers in places where they had no business: roadsides, streams, beaches, and the ocean. The number of public places that provide receptacles for recycling plastic bottles is still disturbingly low. An estimated one-third of PET bottles are recycled, and many of those collected are shipped to China for reprocessing into fiber—an American export item to China! Now the bad press—including questions about bottled water quality and unethical extraction practices—has led to a backlash. Bottled-water consumption peaked at 29 gallons per person in 2007. In the last two years, consumption rates have slipped by a few percentage points—as have recycling rates, the greater mystery. But elsewhere in the world, especially the developing world, the bottled-water market continues to grow, in no small part because of poor local water quality in some regions.

In this topsy-turvy world, what cheers the investor brings the environmentalist to tears. From the website of the nation's only dedicated School of Packaging, at Michigan State, comes Coca-Cola's richly Orwellian description of its business. It accompanied the school's announcement of a $400,000 grant from Coke to be used for developing more "sustainable" containers:

> The Coca-Cola Company is the world's largest beverage company, refreshing consumers with more than 450 sparkling and still brands. Along with Coca-Cola®, recognized as the world's most valuable brand, the company's portfolio includes 12 other billion-dollar brands, including Diet Coke®, Fanta®, Sprite®, Coca-Cola Zero®, vitaminwater, POWERade®, Minute Maid® and Georgia® Coffee. Globally, it is the No. 1 provider of sparkling beverages, juices and juice drinks and ready-to-drink teas and coffees. Through the world's largest beverage distribution system, consumers in more than 200 countries enjoy the company's beverages at a rate of 1.5 billion servings a day. With an

enduring commitment to building sustainable communities, the company is focused on initiatives that protect the
environment, conserve resources and enhance the economic
development of the communities where we operate.

With pockets as deep as the Mariana Trench, Coca-Cola liberally
offers sponsorships not only to the Michigan State program but
also to groups like the National Recycling Coalition, the Ocean
Conservancy, and the UN's Global Water Challenge. Its website
glows with warm, fuzzy eco-words: "stewardship," "collaboration," "partnering," and "watershed protection." There's a word
for it: greenwashing. But green verbiage failed to assuage a group
of socially responsible investors pressing for a commitment from
Coke to address questions about its ongoing use of BPA lining
its beverage cans. The company persistently declines to respond
in a curiously impolitic fashion.

Here's another good news/bad news bite. Freedonia, the research firm, projects global demand for "foodservice disposables"
to increase 4.8 percent annually through 2013. This is a sector
worth $48.6 billion a year, with a product that mostly defies post-
use recycling. The United States is by far the biggest user, but
gains here are expected to be slight. As with caps and closures,
real growth will occur in China and other developing regions.
Economic development means more work, less downtime, and a
premium put on convenience over value. As Freedonia put it,
"advances in the quick service restaurant industry" will provide
that convenience. Take-out food stands are the primary users of
these disposables. Freedonia opines that "pressure" will force
wider use of fancier biodegradable disposables (more expensive =
higher market value) and away from polystyrene. Polystyrene, of
course, can be hard (plastic forks, spoons, and knives) or foamed

(insulated clamshells, hot beverage cups). A piece of heartening news is that Starbucks South Korea is aiming for a 30 percent reusable mug rate. Regarding the compostable disposables, one has to pose the question: Will they be composted?

You may think it's wrong to knock large corporations for their green efforts. You may think they will lead the way to a greener way of life. And while greening efforts are to be encouraged no matter who makes them, they will not be genuine until a company begs you to reduce consumption of their product. And that will never happen. Coca-Cola pioneered the use of plastic bottles for carbonated beverages in 1970. Now they boast of delivering 1.5 billion individual beverages a day, worldwide—most in plastic bottles. For any publicly held food, beverage, or consumer goods company, the Darwinian imperative is grow or die. What gets greener is the window dressing.

In a much-heralded move in 1990, McDonald's abandoned foamed polystyrene clamshells manufactured with ozone-depleting chlorofluorocarbons (CFCs). But they still dispense hard polystyrene utensils and all those sauce packets and plastic straws, and still use polystyrene foam, albeit a less toxic form. That said, McDonald's is a smart company that has discovered how to turn green to gold as they serve up cheap food. On their website they plug their new packaging with recycled fiber content, as well as their recycling and composting efforts. In some regions they recycle frying oil to help power the diesel delivery fleet. The overall weight of their food packaging has been reduced by half. It's encouraging, but it wouldn't be happening absent social pressure and marketing benefits.

I decide to drop in on a local McDonald's to see for myself how they serve their food these days. My coffee cup is expanded polystyrene with a paper sheath. The snap-on top is also unrecyclable polystyrene, as are the utensils. The ice cream treats come in clear polypropylene cups with domed tops. These are potentially recyclable, but I am guessing few are. The store does not offer separate bins for plastic, paper, or compostable waste. I am told the breakfast McMuffins still come in a foam clamshell but was assured it was the "good" kind. McDonald's boasts of serving 47 million customers each day at 36,000 restaurants in 229 countries. They are still doing their part to add to the earth's plastic layer (and its customers' adipose layer).

If you check trade e-zines for trends in packaging, several stand out. *Sustainability* is always near the top of the list. I regard the term warily. It sounds good, but companies are prone to loose definitions of *sustainability*, often using it to mask a product's lack of same. The Sustainable Packaging Coalition (SPC) is an offshoot of GreenBlue, an environmental advocacy group founded by William McDonough and Michael Braungart. These two are the well-known authors/conceptualizers of *Cradle to Cradle*, the blueprint for an earth on which no products are toxic and they all wind up, by design, as "upcycled" new products. The SPC boasts an impressive roster of recruits. The nine founding members include Aveda/Estée Lauder, Dow, Cargill (owner of NatureWorks!), Nike, Starbucks, Unilever, the British-Dutch conglomerate with revenues exceeding $50 billion and a brand portfolio that includes Lipton, Bertolli, Knorr, Dove, TREsemmé, Pond's, Ben & Jerry's, Slim-Fast, and a skin-lightening cream called Fair & Lovely, marketed to Indian women as a product that will improve career and romantic prospects. Unilever claims to sell 150 million items a day, which is nearly 55 billion packaged items per year. Other SPC members include Clorox, Microsoft, Bristol-Myers Squibb, Walmart, Target, and two of the largest packaging companies in

the world, which you may not have heard of, M
Amcor and Wisconsin-based Bemis. The latter
1859 as a Midwest maker of machine-sown cl(
and now operates eighty-one factories in thirte
annual sales in the neighborhood of $5 billion. Packaging con...
nies don't offer jars, bottles, boxes, or pouches. They offer "solutions." Always. These companies and many others display the
SPC logo on their websites. SPC has worked up a definition of
sustainable packaging. It is as follows:

- Is beneficial, safe & healthy for individuals and communi-
  ties throughout its life cycle
- Meets market criteria for both performance and cost
- Is sourced, manufactured, transported, and recycled using
  renewable energy
- Optimizes the use of renewable or recycled source
  materials
- Is manufactured using clean production technologies and
  best practices
- Is made from materials healthy in all probable end of life
  scenarios
- Is physically designed to optimize materials and energy
- Is effectively recovered and utilized in biological and/or
  industrial closed-loop cycles.

This is a tall order indeed for publicly traded corporations
that must pay very close attention to their bottom lines and can't
afford investments that don't promise substantial returns. Of
course, these criteria are meant to be aspired to, with the coalition's help, and failure does not mean banishment. It's a very good

.ng to get large multinationals on this track and actually implementing many of the SPC's ideas. Walmart, for example, has given its suppliers checklists and goals for reducing packaging waste as part of its "sustainability index." Most larger companies now have staff toiling away at raising their green quotients. I'd be willing to bet one of my prized Garbage Patch toothbrushes that no corporate website lacks the word *sustainable*. And in a felicitous quirk of fate, corporate greening often means improved profit margins. Here is what Paul Polman, the CEO of Unilever, said of its ambitious plan, announced in November 2010, to "decouple" growth and environmental impact:

> We are already finding that tackling sustainability challenges provides new opportunities for sustainable growth: it creates preference for our brands, builds business with our retail customers, drives our innovation, grows our markets and, in many cases, generates cost savings.

While crediting Unilever for its fine intentions and efforts, it's hard to resist parsing Mr. Polman's statement thus: "Hey, stockholders! This sustainability thing is a marketing bonanza! When our customers hear we're sustainable, they love us even more and buy more of our products. Innovation! Yes! [Innovation is another chime ringer.] Plus—and this is the best part of all—when you use less raw material, energy and water to make thinner packages and containers (and quite possibly fewer human workers because you automate while tweaking the systems), you lower costs and beef up profit margin."

Another popular buzzword is *metrics*. The metrics of corporate sustainability is this: a company will go greener to benefit its image if the bottom line benefits. If you push too hard on the

SPC's final point, watch out. Extended producer responsibility—in which manufacturers take back packaging—is the third rail of corporate sustainability policy. Here's what the packaging section head at Procter & Gamble had to say on the subject when asked in a roundtable discussion organized by *Packaging Digest*:

> Regarding extended producer responsibility . . . [e]very entity in the supply and value chain has a role to play. To simply ask companies to pay for packaging waste isn't going to drive right behavior from consumers. What we really need is a reinforcing mechanism to drive consumers to want to recycle instead of just throwing things away in a landfill. It has to be a holistic approach.

So trash generated by Procter & Gamble is a "people problem." People need to be coaxed toward "right [recycling] behavior" by kindly paternal corporations. Tough love is what's needed, but there's a hitch: with new variations on compostability, recyclability, and bioplastics that may or may not be compostable or recyclable, what does the consumer do? David Luttenberger, a packaging vice president at Iconoculture, a Minneapolis-based global consultancy firm, addresses this issue in *PL* (*Private Label*) *Buyer*: "Consumers are still wrestling with paper versus plastic, and now we're throwing all these claims at them like degradability and compostability." What are likely results of this confusion? More plastics headed for the landfill and to roadsides, beaches, urban rivers, and the ocean, where cool, wet conditions put the brakes on degradability.

Some in the corporate world thought and probably hoped sustainability was a passing consumer whim. The director of packaging innovation and development for Sara Lee has been quoted as saying: "A lot of people thought sustainability would go the way of RFID [radio frequency identification, technology used mainly by Walmart to track inventory] or be killed by the recession." *Packaging News*, a UK trade magazine, moans in a trends article titled "The Never-ending Story": "It's the issue that won't go away: sustainability. And the year brought a constant stream of new angles from brands and retailers . . ." But if plastics are the least recyclable and most environmentally problematic of packaging materials, why are more R&D dollars spent on developing them, according to Freedonia, and why does the post-use problem worsen?

One must state the obvious. Glass provides the same "barrier protection," but it's much heavier, and it breaks. Paper only works for dry or frozen goods, and paper used with frozen foods is typically sheathed in polyethylene. Also, paper is opaque. (Cellophane can be considered a bridge material between plastics and paper, being derived from chemically treated natural fibers that biodegrade.) Metal costs more and is comparatively scarce. Plastics are cheap and astoundingly versatile. Without them the planet would look different, and so would the economy.

Freedonia predicts worldwide demand for plastic packaging to grow from 10.8 billion pounds in 2009 to 12.4 billion by 2014, an increase of about 20 percent. The figures are numbingly vast. This growth, it says, will be fueled in part by new tricks: treated plastics prolonging shelf life, "resealability," and "microwavability." Protean plastics simply outperform paper and other options. But it's rarely either/or these days. Larger electronics come in rigid cardboard boxes (plastics can't do that with the same weight) braced inside by formed plastic foam and swathed in polyethylene film. Metalized plastic film is an ingenious newer scourge. I often see crinkled tear strips from this packaging type in shore waters—from snack-sized Doritos and Cheetos as well as "wholesome natural" bars.

The exigencies of modern life make hypocrites of us all. When food gets shipped from one part of the country to another, it's not only about light weight and freshness but about spoilage and food safety. According to the European Plastics Industry, only 5 percent spoilage loss from "farm to fork" occurs there, attributable to plastic packaging. In the developing world it is more like 50 percent. It saddens me to see how hygienics are used to justify the rapid introduction of plastic packaging to people who traditionally used fully biodegradable packaging made from plant material and who lack disposal infrastructure. We see rivers in East Asia looking like the trash islands people envision in the gyres. If disposable plastics continue streaming into less developed regions, the result could well be those heretofore mythological plastic islands in the ocean's gyres. David Osborn of the United Nations Environment Programme spoke at a marine litter workshop at the European Commission in Brussels where I presented. He suggested plastic packaging have labeling similar to that on cigarettes in the United States, warning of threats posed by plastic litter to wildlife: suffocation, starvation, blockages.

The faux green tour de force might be the Swedish-born Tetra Pak. Until very recently, Tetra Pak was the largest packaging company in the world. (Amcor bought half of Alcan in 2010, becoming the biggest.) Its inventor, Ruben Rausing, died in 1983, the richest man in Sweden. But Tetra Pak is still impressive, being essentially a niche product. On one hand, this container-package in its several guises is a miracle of materials engineering. It's the ultimate hybrid: six super-thin layers of low-density polyethylene, paper, and aluminum foil that are laminated, sterilized in a fog of hydrogen peroxide, dried, cut, folded, filled *from the bottom* with liquids flash pasteurized (in the case of milk and other products prone to contamination) using UHT ("ultra high temperature") technology that kills pathogens but, the company claims, preserves nutrients. The rectangular cartons are packed into shipping cases with no space wasted—a soup can or wine bottle can't make the same claim—and are lightweight, airtight, and protectively opaque. The format allows shelf storage of a year or more. At natural-food stores—wonderlands of eco-cognitive dissonance—various non-dairy milks (soy, rice, almond, oat) as well as organic soups come in Tetra Paks. The famous "juice box" is a Tetra Pak and a common beach cleanup item, as is its little detachable plastic straw. Now wine in Tetra Paks is catching on, and these screw-off plastic caps have become beach litter. Twenty-two billion of these containers are produced each year.

The company's marketing material makes extravagant claims of sustainability. The paper is sourced whenever possible from managed forests. The product-to-packaging ratio is the best in the business, 96 percent, better than PET bottles, and leaves glass in the dust. And it's recyclable. Yes, Tetra Pak has a special technology that pulps the cartons and separates the different materials so the paper can be turned into toilet paper. Only this technology is not readily available: Florida is home to the only facility in the United States. During a recent Tetra Pak "Twitter conference" on sustainability, one "attendee" asked what percentage of Tetra Paks is recycled. Oddly, given the conference theme, he was tweeted that this data was not available. Elsewhere, a recycling rate of 30 percent is claimed for Europe and low double digits for the United States. Twenty-five states reportedly offer Tetra Pak recycling. The situation in Canada is said to be somewhat better. On the bright side, Tetra Pak says its cartons take up less space than competing containers in landfills. And yet, as a Toronto-based TreeHugger columnist put it: "In what world can you call Tetra Pak green?" As to wine in Tetra Paks, he argues the only true green "solution" is to wash and refill existing wine bottles.

A special planetary risk of these packages is their tendency to float out to sea, then become waterlogged and sink to the seabed — another fragile, imperiled habitat where the vital, natural, environmentally stabilizing process of gas exchange occurs.

A company that includes a small amount of recycled material in its packaging touts its green credentials. But so do the lucky companies that have always packaged their products in paper, metal, and glass, the most easily recycled materials. The Freedonia Group says green packaging is where the growth is and will soon be worth $41.7 billion. But they also say that the definition of *green packaging* is so broad that "virtually all companies offer packaging that can be considered green." Whatever: plastics are winning and are predicted to overtake paper as the reigning packaging material by 2014. If a company switches to thinner-gauge plastic for its jam bottle, they'd like you to think it's sustainable, a guilt-free purchase. But if you'd rather sustain the planet than a global conglomerate, simply buy fewer packaged products; hold on to things, like computers, for as long as they work; get as much food as possible from a local farmers' market or your own garden. And eat less fish, because fishermen still dump way too much stuff into oceans and are too good at what they do: catching fish.

At times, even the experts seem conflicted. Some forecasts warn of rising petroleum costs and concede that "many consumers think plastic has poor environmental credentials." The forecast group Visiongain appears to be in this camp. Based on paper's stronger environmental bona fides, the London-based firm ventures that claims of plastic dominance may be overrated. But while experts may lack consensus, the overriding reality is clear: churning out unquantifiable tons of plastics that lack "end-of-life planning" invites eco-mayhem.

Trend lists and corporation websites never fail to invoke sustainability's buzzword twin: *innovation*. As a connoisseur of consumerist excess, and as one who is ever on the lookout for the next new exotic plastic trash species—and as a fruit grower and blender owner—it was with a mix of dread and delight I came upon this breakthrough:

> Consumers are embracing the smoothie trend in a big way. . . . Recognizing that many produce shoppers would like to consume smoothies more often, but find them too difficult to make at home, Del Monte brand has introduced a line of ready-to-blend Fruit Smoothie kits. The all-in-one kits contain fruit chunks and puree. Consumers simply add ice, blend and enjoy.

Smoothies—the all-natural fruit shake invented by health-conscious hippies in the sixties—then co-opted by a franchise empire, now in a plastic kit on your supermarket shelf. It's an innovation and a solution—unless the idea of more processed foods in plastic packaging makes you cringe.

We stake the health of our economy on "innovation," assuming it is always good. But when innovation leads to 26,893 new packaged foods and products—often in or of plastic—in one year, 2009, it's time to slow down and consider the Pandora's box of consequences of all this new-and-improvedness. We need to stop cultivating innovation for its own sake and start thinking morally and ecologically about the innovations we embrace. Is it worth trashing the planet to grab the newest, coolest, most convenient thing that comes along? As far as I'm concerned, each purchase should be a moral decision that takes into account the life cycle of all the materials in your shopping basket, both their origins and their fate.

constant shift of burden

demonstrating harm

POP's persistent organic pollutants
  – bioaccumulation

CHAPTER 9

endocrine disruptors : affect hormones

GONZO SCIENCE

THIS DATA FROM THE GYRE VOYAGE feels like a ticking time bomb. Each new day could be a day on which a few people, or many, learn their ocean is turning into plastic stew. If they knew, they might begin to see plastics in a new way, maybe handle them more carefully, maybe begin to make different choices. But there's a process involved in getting the word out, and it cannot be denied. Scientific credibility and the authority that comes with it is key. So, in spring of 2000, I'm still drafting my first scientific paper and also finishing a project with colleagues in Baja California, in which we compare eastern Pacific gray whale calving lagoons. One is disrupted by a salt extraction operation and the other is unspoiled. We are gratified when the Mexican government and the Mitsubishi Corporation announce that they will abandon plans to construct a massive salt works in the undeveloped lagoon. It's around this time that word comes of a conference that seems too important to miss. Here's another chance to hone our message and connect with established marine debris experts. Maybe a few will be sufficiently intrigued to help investigate the impact of billions of plastic bits suspended in north-central Pacific waters.

The event is the Fourth International Marine Debris Conference on Derelict Fishing Gear and the Ocean Environment. It's scheduled for early August 2000 in Honolulu, six years after the third conference, held in Miami. Out of the Miami conference came a book, *Marine Debris: Sources, Impacts, and Solutions*, edited by James Coe and Donald Rogers. This may sound a little dry to the average reader, but it became my virtual Bible for a time. I see Coe on the roster for this conference, as well as other "stars" in the field of marine debris. This seems too good an opportunity to pass up, especially now that we have a firmer grip on harm—a well-defined context if not proof. The deadline for presenting a poster has passed, but there's still time to submit a research abstract. This is simply a way of getting our research area into conference literature. Susan Zoske signs up both herself and me as attendees. But it rankles. Missing the poster deadline is seriously disappointing. Ours is the type of research that would lend itself perfectly to this graphic mode of presentation. Not one to be deterred, I devise a plan for carrying out a guerrilla maneuver.

And I get to work assembling a crew for another voyage to Honolulu. Recruits include Emiko Kobayashi, a Surfrider activist and environmental studies major, and Javier Santiago Acosta, a marine science graduate student at the Autonomous University of Baja California. I'd worked with Javier on the Bight '98 study and the recent lagoon project, and I'd noticed how well he handled himself on the *Alguita*. We set sail on July 20. The first day of the conference is August 9. The crew is rounded out by two veterans from the first research voyage: my neighbor Mike Baker, who'd served with the California Highway Patrol, and the birder, Robb Hamilton.

On this 2,500-mile voyage, we take the normal outbound sailing route, basically a rhumb (straight) line heading southwest connecting with easterly trades due west of Baja California. This is an efficient plan designed to get us to the conference on time. We'll bypass the high where debris lurks and winds are weak. We make good time. Kobayashi is sidelined by seasickness but bounces back after a few days.

We can't resist doing science on the fly and drop trawl nets randomly along the way. Not one is free of plastic. Mind you, this is well south of the convergence zone. It could be that these unexpectedly productive trawls will provide a clue to the route debris takes to the garbage patch. Later we'll analyze the samples and add the results to our growing database. This time we'll streamline the process by comparing planktonic mass with the plastics—but not counting individual planktonic organisms as we did before. The results will later show a higher plastics-to-plankton ratio than we expected, but predictably lower than in the gyre. At the end of each trawl, we heave to and make dives to visually confirm that what we're trawling fairly represents plastics' ubiquity out here. I give the crew small aquarium nets to swipe through the water so they can capture samples of what they see. They're always there, the resin bits, fluttering like snowflakes in the water column. I start a notebook. On each dive I use my aquarium net to catch subsurface plastic fragments. I tape these flecks into an old accounting notebook my father had in the 1940s and make a written record for each sample, noting the location and conditions of its capture. Our log for the voyage also mentions deck observations of drifting packing straps, crates, and fishing line, and consumer items such as soap and deodorant bottles.

Susan Zoske flies in from Los Angeles and meets us at our Honolulu slip. The crew scatters, eager to unwind in Honolulu. The next day she and I rendezvous at Honolulu's sleek new Hawai'i Convention Center and register without anyone looking askance at the substantial unsanctioned poster tucked beneath my arm. We repair to the exhibition area outside the main auditorium where others are setting up their displays. As an aside to readers who've yet to attend a science conference, it's not unlike your standard middle school science fair, only what were once spoken reports with a few slides, and are now invariably PowerPoint presentations. Months beforehand, organizers issue requests for research papers to be considered for the official program. Would-be speakers whose talks can't be squeezed into the schedule—and others who may be inclined to do so anyway (often career-minded graduate students staking out new areas of research)—are invited to show posters that graphically present their findings. During breaks in the proceedings, attendees stroll through the exhibition area. Poster presenters are usually standing by, poised to answer questions, hand out leaflets, and network.

Scientific posters can be straightforward or fancy affairs, usually in triptych format combining text, graphs, and images. Stylistically, ours is somewhere in the middle. In the weeks before setting sail, I could be found hunched over our dining room table, glue stick in hand, paper cutter nearby, affixing graphs and tables prepared from our data to a triptych-style craft store poster board. I printed out a slightly jazzed-up version of our study's abstract and glued it in the middle. I also affixed actual plastic petri dishes containing size-classed plastic flakes from one of our trawls. I figure showing is always better than telling, and this way people will be able to connect material summarized in the graphs and tables with the real deal. We thought it turned out rather well.

In the exhibition hall, the posters are set up on easels and long tables in a room the size of a basketball court. It looks like there may be about fifteen, all told. We spot an idle easel and grab it, deciding furtive won't work and acting like we belong. Many times poster presenters are late or fail to show, and it appears this may be the case at this event. We spot several feet of unoccupied space sandwiched between other posters, and make a beeline. Setting up, we half expect a stern organizer to approach, clipboard in hand, check our poster against a master list, then summon security. But this is Hawaii, land of aloha and the spirit of inclusion. Our stealth tactics seem to work. Either we blend in, or organizers are simply too busy to question our legitimacy— probably both. Zoske has prepared a stack of packets filled with copies of my draft article's abstract, the data graphs, information about Algalita, and a sheaf of newspaper clippings. Science may have well-fortified gates, but the local press from Santa Barbara to San Diego, including the *Los Angeles Times*, has embraced the story of what's becoming widely known as the Great Pacific Garbage Patch, and my adventures there. Why, one could ask, even bother with scientific credibility if free publicity is there for the plucking? There is a good reason. Science—and not always the best science—determines policy. Science plus strong public sentiment are even more powerful drivers of policy shift. If you can also harness law, when policy can be proven to do harm or lack enforcement, you have a recipe for change. We've seen this happen time and again, with unleaded gasoline and paint, with cigarettes, and with toxic synthetic chemicals like DDT and PCBs.

We are approached by the woman who is cochair of the exhibitor/poster committee, Kathy Cousins. We are fortunate indeed. She instantly assuages any residual trepidation. Far from showing us the door, she reveals herself as an instant ally. It could be said she gives us, here in the early moments of the conference, a false sense that the entire experience will be a lovefest. Cousins at this time is a wildlife biologist with the National Marine Fisheries Service, a branch of the conference sponsor, NOAA. She spends time on Midway Island, nearly twelve hundred miles northwest of Honolulu, where Laysan albatross chicks by the tens of thousands perish each year, stuffed by their well-meaning parents with plastic non-food mistakenly culled from the pelagic food court. Policy dictates that the nesting area is off limits during the long breeding season, meaning it can't be cleared of accumulated plastic trash. Not only is plastic debris flown to the remote island by the albatross but it's spat out, as Ebbesmeyer would say, by the debris-laden gyre. It's an alien invader in the faraway archipelago that includes Midway and other ecologically important patches of land that for eons were clean and natural places. Cousins tells us she's seen, firsthand, evidence of a correlation between plastics, which often line albatross nests, and thinning of seabird eggshells. This suspicion of toxicity is both intriguing and disturbing. It seems to support my unshakable but inchoate sense that plastic has yet to fully reveal the harm it's capable of. We figure it's partly the stress of working on this major conference, but Cousins begins to cry as she expresses her sorrow and frustration. I'm not surprised when, a few years later, I learn she's moved to Idaho, where she now works with the U.S. Fish and Wildlife Service on habitat protection.

I meet several people whose work I've studied and experience the great feeling of satisfaction that comes from finally interacting

with people whose intellects you have admired from afar. But if
the lesson learned at the San Diego symposium was the need to
show harm, the takeaway from this event is this: to the derelict
fishing gear crowd, size matters. Microplastics seem, literally, of
small concern to them. They're all about massive clumps of plastic net, monofilament line, floats, and buoys. That's probably
because the marine debris arm of NOAA is under its Office of
Response and Restoration, the agency that deals with shipwrecks
and their impacts on sensitive habitats. I feel like I'm peering out
from a diving bell, exploring new underwater terrain. And from
this vantage point, it's also clear that plastic, the material, is of far
less concern than the derelict nets and other types of fishing gear
made from it. I wonder how the two can be separated. "Marine
debris"—defined as man-made solids adrift or abandoned in the
oceans and Great Lakes—was a non-issue until plastic fishing
gear came on the scene starting in the 1960s. It's the special durability and buoyancy of plastic that makes derelict plastic fishing
gear such an environmental nightmare and a worthy subject for
an entire international conference.

I also learn *activist* is a label that requires judicious deployment. It's almost visible—the line drawn between scientists who
study problems and people who agitate for solutions. Frustrated
global warming scientists are breaking down the distinction, but
political reactions to their warnings expose the reputational risks
of scientist activism. With some, like Kathy Cousins, activism
forges a bond. With others, you best be all business or expect a
very cold shoulder. It actually seems a little personal with some
of these guys. Which also tells me something about why the issue
of plastic debris isn't stirring concern outside of government
agencies and academia. I learn that, in fact, many of these
researchers do have a sense of mission. But "professionalism"

breeds understatement. For those at this conference, the end goal appears to be policy change within the narrow confines of the fishing and shipping industries and the agencies regulating them. This is needed, to be sure, but what about the umbrella handles, the soda bottles, the cigarette lighters, the shoes, and the soccer balls we're seeing out there? Consumer products. I'm thinking a lot of this stuff is coming from land, not fishing boats, and that it could be as malignant in its way as lost nets. I also know that much of what's out there are these very things—the nets, the floats, the line—broken down over time into billions and trillions of tiny pieces. What's the impact of these microplastics, which surely represent the fate of all the big plastic derelict fishing gear that eludes removal?

I lunch with a group that includes James Coe, coeditor of the marine debris book that's proved so helpful. But I get no traction, to put it nicely, talking about the ocean's plastic fragments. Among plastic debris research stars, Dr. Anthony Andrady shines brightly. I have no idea who he is when he stops by the booth—a small man with an intriguing accent—but jump a little when he introduces himself. He's Sri Lankan and known to all in the field for his groundbreaking research into how plastics degrade and for editing the definitive text *Plastics and the Environment*. He's here from the Research Triangle Institute International, a global R&D firm based in North Carolina. At first it's clear he's thinking we're anti-plastic activists associated with Greenpeace and wanting to ban all plastics. I assure him that banning all plastics is not our mission and turn the conversation to how we might work together. I tell him that he is the general and we are the soldiers, awaiting his orders. If we're to clean up the oceans, we need to know what the plastic fragments used to be and how long they've been there, and we need to get an idea of where they're coming from to stop them at the source. Microplastics pose a particular challenge. Most would be legacies of objects discarded long ago and broken down over years, likely decades. But we need to know the proportions coming from land and ship. The prevailing sentiment at the conference seems to be that most marine debris is ship sourced, but to this day the source issue bedevils researchers. Andrady offers to analyze samples from this most recent crossing, and I send him a small Baggie's worth within a month. He confirms what we thought: the particles are degraded polyethylene and polypropylene. But beyond that, it's anyone's guess what they once were and where they came from.

I attend a lecture by James Ingraham, Curt Ebbesmeyer's partner and developer of the OSCURS model that predicted the garbage patch. In the 1970s he "seeded" areas of the North Pacific, from Southeast Asia to the Bering Sea, with buoys and tracked them for twelve years, long enough for two turns around the entire North Pacific Gyre. There were two areas in the North Pacific where most of the buoys tended to migrate: my new bailiwick, the Northeast Subtropical Gyre, and its counterpart to the west between Hawaii and Japan. His big point is that efforts to locate and remove derelict gear should be focused in these areas. There are talks about tens of thousands of northern fur seals being killed by abandoned nets. And about the risks entailed in peeling seventy-seven thousand pounds of net from the delicate coral reefs north of Hawaii. And about fishermen who have little incentive to responsibly handle their gear and trash due to weak or nonexistent collection systems in most ports, disposal fees, and an anything-goes attitude shared by many who make their living on a very large, unpoliced ocean. MARPOL Annex V appears to be a toothless treaty. A Taiwan researcher bracingly admits that *all* Taiwanese fishermen jettison their nets at sea after catching all they can. The space the nets take up is more profitably occupied by catch. The topics discussed are fascinating, but none make mention of the inevitability of big fishing gear breaking into billions of small bits. Or the likelihood of these bits becoming dinner for the food chain's lower members, the filter feeders.

At the end, "solutions" workshops are scheduled. I choose group D, Industry. It looks to be a smaller forum, meaning I may have a better shot at being heard, but it's also what I want to talk about. I plan to present my big idea for addressing the debris problem.

In the systematic and well-meaning way of facilitators intent on emerging from conferences with an impressive document pointing the way to an improved planet, a list is made on a large whiteboard and notes taken on a Brobdingnagian pad of paper hanging from an easel. These are phrases I hear: *More research needed. Funding sources needed. Interagency partnering. High-tech monitoring. Awareness. Education.* I make my move, standing up and suggesting that we work directly with the plastics industry. After all, it's their material that's causing the problems. Shouldn't they be included in discussions of ways to fix it— perhaps take some responsibility, even financially?

I am stared at, and the facilitator refuses to add this suggestion to the list for further "prioritizing." They say the plastics industry is "off topic." Excuse me, I say. Isn't industry the focus of this session? And is not the plastics industry an industry that is relevant to this forum? They say no, it's about the fishing industry. Then participants are given stickers and asked to affix them to the list of suggestions as a way of ranking perceived importance. I'm beginning to feel I'm at a Scout meeting. I put my sticker on my forehead. What I have to say is important. I am stared at again.

No one seems to want to go to the heart of the problem. It's the plastic material itself that is formed into almost every debris thing out there: the massive trawl nets, longline by the mile, floats and buoys, and other items commonly tossed overboard by fishermen: bleach bottles, light sticks, butane lighters, plastic crates, and empty plastic chemical drums, which have the curious property of drawing fish to them and are known as "fish aggregating devices," or FADs. My suggestions will not be written in felt pen on the very large pad, nor will they be included in the published recommendations. This will not be the first time I hear that plastics pose a "handling" problem—that is, a "people problem," not a material problem. It's all our fault, in other words, and fisherman are just seafaring litterbugs. I leave this session in disgust and go to another one, but it's in its final stages.

A few months later I receive the conference proceedings, the record of talks given, posters shown, and recommendations made. The planning session write-ups brim with excellent ideas, and also impediments and costs of implementation. But I am still thinking that responsibility and accountability are spread too thin among agencies and stakeholders, and that three previous conferences have apparently done little to stem the marine debris problem. I'd be willing to bet any amount that by the time the next conference rolls around, there will be more, not less, plastic in the ocean—for small marine creatures to ingest at their significant if yet-to-be-scientifically-proven peril. I do not see evidence that a connection's been made between derelict fishing gear, 95 percent of which is made of plastic, and the mid-ocean blizzard of plastic bits. Neither am I convinced that narrowly addressing the derelict fishing gear problem—which, for the near term, can only be marginally mitigated, given the regressive practices of rogue fishing nations, artisanal fishermen, and the impossibility of enforcement on the high seas—is the way to a cleaner ocean.

The proceedings include a section about the poster session, in which the posters are classified by category and listed with descriptive passages. The three official poster categories are "monitoring, enforcement, and removal," "ocean stewardship, education, and outreach," and "debris prevention and legal issues." At the end of the poster section is a fourth and final category, and it's mine alone: "Other." Okay. I've been taught another lesson or two. But guess what? I'm more determined than ever. And I console myself knowing we're inching our way into "the conversation," though it could be viewed as crashing the party. Best of all, we've made a few priceless connections.

ELEVEN YEARS LATER, in March 2011, history repeats itself, with a few wrinkles. NOAA has organized a Fifth International Marine Debris Conference in Honolulu, this one with the theme "Solutions." The very first conference was held in 1984, so I'm glad to know it's finally time to talk about solutions. After all, the oceans' plastic burden has grown exponentially in the last twenty-seven years. This time Algalita makes the poster deadline and in fact sends a delegation of six scientists, armed with posters and PowerPoint presentations. My talks are on citizen science and my ten trips to the gyre. We take orders for foldable, portable aluminum manta trawls designed by Algalita's engineering and heliarc welding whiz, Marcus Eriksen, and for our plastics identification field kit, developed with Anthony Andrady, whom I'd met here eleven years earlier.

But some things are unchanged. At the well-attended plenary session, Algalita's game-changing efforts are acknowledged by

environmental delegates from the European Union and the UN
and by assorted others, even a Coca-Cola representative. But I
still get no love, or passing mention, from NOAA. Maybe the
sticker incident still stings. A Korean film crew follows me around
and I find myself the "esteemed guest" at fund-raisers, debris art
openings, and other side events. A far cry from 2000, and more
fun. But a chasm has widened, one that at the 2000 conference was
a mere hairline crack. Of this conference, which attracted more
than four hundred attendees, twice the number at the 2000 event,
Roz Savage, the intrepid Englishwoman who is rowing solo
around the world, will make a striking and apt observation. She
writes in her blog shortly thereafter: "I had thought that plastic
pollution was much less contentious than, say, climate change,
but it seems that there is no limit to humankind's ability to find
grounds for division rather than cooperation."

Conference sponsors include Coca-Cola and the American
Chemistry Council, and they have seats at the table, behind
closed doors, where government agency officials are drafting the
"results-oriented" Honolulu Strategy, a plan to lead the way to
a debris-free ocean. Only word leaks out that this document,
incredibly, will omit the word *plastics*. When this comes to light,
Dianna Cohen, a founder of the Plastic Pollution Coalition and
a strong Algalita ally, rallies opposition and gets results. Normally reserved academics, including the University of Hawai'i
at Mānoa's Nikolai Maximenko, who is mapping debris plumes
from the 2011 Japan tsunami, as well as outspoken ones, like
Richard Thompson, a noted microdebris researcher and professor
at the University of Plymouth, call on government to confront
industry and put responsibility where it belongs—on plastics
producers—not volunteer cleanup crews, taxpayer-supported
government agencies, and NGOs. Source control is glaringly

the least focused-on strategy, since it affects industry's bottom line. And, as I tell a reporter from Kauai, all the creative and well-meaning schemes focused on cleaning up the gyre, will work about as well as trying to bail out a bathtub with the tap still running.

NORTH PACIFIC CENTRAL GYRE, Lat. 38 56'N, Lon. 142 37'W, Thursday, September 7, 2000, around midday. We're en route back to the West Coast from the first Honolulu conference, sailing under clear skies in calm conditions. We have new crew members. One is Commander Daniel Whiting, one of the U.S. Coast Guard's few oceanographers. He's now retired and glad, he says, to be doing "real" science of a type not practiced in the Coast Guard. This voyage also represents adventure to him, the type he dreamed of when he graduated from high school and enlisted. It is he who dubs our research methodology "gonzo science." Another crew member is Tony Nichols, a volunteer whale-watching guide with the American Cetacean Society. He anoints me "the debris avenger." Javier Santiago Acosta rejoins us for the cruise home. We also welcome Chris Thompson, a commercial organic farmer in Santa Barbara who added fine cuisine ingredients to the 1999 cruise with crates of fresh organic produce. His concern for the planet is large, but his experience on the high seas is minimal. What he lacks in seafaring experience he makes up for in spirit and people skills.

It's day 13 at sea. We'd weighed anchor at Hanalei Bay, on the ruggedly beautiful north shore of Kauai. On this return leg, we've plunged back into the gyre. The sea state is 2, the wind at about five knots, the ocean surface nearly glassy calm. We pull up a ghost net with the gantry and photograph it, but it is too heavily fouled with barnacles to lift on deck. The 1400 log entry reads: "Collecting mucho debris." Then we begin to notice something highly out of the ordinary even in this extraordinary garbage patch. One of us spots a crisp-looking plastic bag nonchalantly drifting by. And then another bag. And then another. And then we realize they're everywhere. We're in the middle of the ocean, surrounded by a sea of plastic shopping bags. It's as if a tornado tore the roof off a floating shopping mall just beyond the horizon line. But these bags clearly burst from a lost shipping container. We see super-large plastic "mother bags," undulating like freak monster jellyfish in the swarm of baby jelly bags. These would be the bags that bagged the shopping bags, all destined for a who's who of North American retailers. We start snagging them and read off the names: Sears, Bristol Farms, The Baby Super Store, El Pollo Loco, Fred Meyer, and the bag champ in sheer numbers: Taco Bell! Chalupa!

Most are your basic T-shirt bag, distinguished by tissue-thin gauge and convenient handle holes, invented in the early 1960s by a Swedish engineer named Sten Gustaf Thulin. This man looked at a long, flat tube made of plastic film and saw a way to cut and weld a bottom seam and die-cut handles at the top. But it wasn't until the late 1970s that this cheaper alternative to the paper bag took hold in the United States. Thirty years on, it's racking up those amazing distribution numbers we see today—possibly a trillion a year. Because the bags skimming by us look reasonably fresh—no fouling by algae, few tears or signs of stress, still stretchy—we figure the spill occurred recently. They were probably blown and printed in Asia and on their way to a store near you when disaster struck, perhaps in the form of a rogue monster wave. Which would be a little odd in these comparatively calm spring and summer months. In any case, we marvel at the small print stating: "Made in U.S.A." Well, it could be one of those technical truths, if the resin pellets were American made. We wonder if we've discovered the *Exxon Valdez* of plastic bag spills, right in the center of the most remote portion of the Pacific Ocean. But wait. Could these bags be flotsam from the "weather bomb" Ebbesmeyer told us about a year earlier, before the '99 research voyage? No, the bags look too fresh. And yet, we've just learned at the conference that Dr. Andrady's research shows a substantial slowing of the degradation process in the marine environment.

We recover about a dozen bags, scooping them into nets or snagging them with grappling hooks as they tumble by. We'd have caught more, but most are just out of reach and the sun is now low in the sky, making heaving to and giving chase with the dinghy a dicey proposition. But, for the sake of science, we make a single three-minute visual survey, looking only off port side and out to seventy meters. We document forty-nine plastic bags across a three-mile transect. Bags drift by for another ten or so miles, and then it's night.

Back on dry land, I ask our maritime law adviser, attorney James Ackerman, if he knows of a detective who might be able to track down the vessel responsible for the bag spill. And, true to form, Susan Zoske goes into pit bull mode when we tell her of this event, serially calling the corporate headquarters of every company represented in our haul. None she reaches is aware of any undelivered bags, which is not so surprising when you consider they were most likely on their way to central distribution centers. The detective recommended by Ackerman thinks he can produce results, but he wants a $5,000 retainer up front. We have to pass. Curtis Ebbesmeyer has shown wizardly skills at extracting information about corporate container spills, after years of practice. Most famously, he went directly to Nike to get information about flotillas of athletic shoes washing up on Oregon beaches in the early 1990s. By invoking the noble cause of science, he sometimes makes headway. But he admits to hitting brick walls as well.

Container losses have big impacts—not only environmental, but navigational. Many are filled with buoyant cargo and don't sink. Tightly sealed, they float near the sea's surface like insensate Moby Dicks, ready to hole and sink the boats of unsuspecting mariners. And yet, despite concerted international efforts by many, including Algalita, companies are not legally required to report such spills. In fact, if the cargo is considered "non-toxic," the ship owner escapes liability for any cleanup. An outrageous and egregious container spill case began to unfold in March 1997. The MV CITA, a German-owned container ship, ran aground off Newfoundland Point in the Isles of Scilly, twenty-eight miles off the southwest tip of England. Coastal residents reported seeing twelve containers floating in the water, while several others washed up onshore. One of the containers had held 1,500 miles of polyethylene film, longer than the U.S. West Coast from the Mexican to the Canadian border. A decade later polyfilm scraps were still washing up on shores throughout the British Isles. The cleanup cost local authorities 100,000 pounds ($250,000) so they duly filed suit in the German court system to recoup the costs from the ship's owner. In 2005, when the case was finally heard, the ruling was not favorable to the British plaintiffs. The verdict, based on international maritime law, found the ship owner immune from liability and—adding insult to injury—ordered the plaintiffs to pay their own court costs. Spill matters are kept private among shippers, customers, and their insurance agents.

All are quite content to hide behind the veil of arcane maritime law. International law conventions periodically try to break their grip, with small success.

Back in port, we submit the plastic/plankton paper to the *Marine Pollution Bulletin* and cross our fingers. Preliminary acceptance, should it occur, means the paper will spend the better part of a year undergoing the rigorous peer-review process, and success is not assured. *Alguita* limps into dry dock for repairs to virtually every system. I ask crew members to write up their impressions of the voyage for inclusion in the *ORV Alguita News*. I am amused to read in Commander Dan Whiting's report the following: "The *ORV Alguita* is not the glamorous, sexy science of made-for-TV hunts for sunken ships. This is the real thing, science in the raw: smelly, dangerous and very demanding."

With so much coverage in the local press, my schedule is starting to fill with speaking engagements. People want to hear about the dark side of plastic, how it's escaped from civilization and colonized the mid-ocean. In sailor parlance, I sense a change in wind speed, a freshening. Steve Weisberg of SCCWRP wants a sample jar from the latest voyage. I knock on his office door and make a personal delivery. That nasty sample still sits there in the same spot on his bookshelf. We talk about designing another study that will either support the 1999 gyre findings or establish them as an anomaly. We plan to take near-shore trawls, where ocean waters are far richer in nutrients and plankton than in the mid-ocean desert found in the gyre—meaning a high plastic-to-plankton ratio was somewhat predictable in a place where plankton is scarce and debris is known to accumulate. Weisberg warns me to expect to hear this very complaint from critics. I'm not bothered, since it's my firm conviction that any plastic at all, let alone thousands of tons of it, has absolutely no place in the middle of the subtropical North Pacific. But near-shore waters may contain more plastics overall, given their proximity to civilization and its plastic excretions. We will see. After addressing the local chapter of the American Cetacean Society at the Cabrillo Marine Aquarium and—to my pleasant surprise—receiving my first honorarium, I'm invited to speak at the society's biennial conference to be held in Monterey. I've given plenty of local talks about marine ecology and organic farming. But this will be a first—a major conference at which I speak as an authority on plastic trash in the ocean.

And another good thing will come of it. My talk will lead to a renewed relationship of momentous significance.

CHAPTER 10

THE MESSAGE
FINDS ITS MEDIUM

IT'S NOVEMBER 2000 and I'm driving up California's winding
Pacific Coast Highway toward Monterey, site of the biennial
American Cetacean Society (ACS) conference, Whales 2000. I'm
no stranger to speaking about marine matters, but this will be my
debut as a plastic debris foe with a special interest in microplastics. I've worked on various projects with Diane Hustead, conference cochair and an officer in the local ACS chapter. She's
proposed me as a speaker because whales and other marine mammals frequently, sometimes lethally, encounter plastic debris in
their habitats. I've long admired the work of this society, the very
first whale advocacy group when it was formed in 1967. Now it
has seven active West Coast chapters and an international reputation. Its core mission is unswerving opposition to the hunting and
killing of marine mammals for commercial gain. Not only cetaceans (whales, dolphins, porpoises, narwhals, belugas, and killer
whales) but pinnipeds: "fin-footed" seals, sea lions, and walruses.

They've called out the International Whaling Commission for accepting luxury travel from Japanese whaling lobbyists and have even gone after Inuit who were reported to be selling whale meat caught under the pretext of subsistence hunting.

I'm one of twenty-one speakers, some coming from as far as Alaska and New Zealand. I've prepared a PowerPoint presentation playing up the angle that should naturally engage this group: the likelihood that the baleen whale diet of small fry food staples—plankton, krill, and small fish—is liberally seasoned with plastic. It's a widely noted irony that such enormous creatures—including the largest animal ever to live on earth, the blue whale—subsist on the smallest. What troubles me is that whale feeding mostly happens near the sea surface, where plastic fragments mingle with and mimic legitimate organisms. The larger species of baleen whale have mouths the size of carports. The baleen itself is a food-filtering structure that spans the whale's upper jaw and looks something like a hair comb. It's made of flexible keratin, a fibrous protein and natural polymer that we find in our hair and nails. Meanwhile, their "toothed" cetaceous counterparts, including dolphins and killer whales, could very well be consuming fish with plastics-spiked diets. Net entanglement has long been a concern of the society, but several recent necropsies of "stranded" (code for dead or dying on or near the shore) toothed whales have found shocking quantities of plastic bags and net fragments in their guts. This makes ingestion an emerging concern. The bag finding is significant. These are consumer or "user" plastics, not derelict fishing gear or industrial pellets. I find that preparing my talk clarifies my thinking on how best to present the issue and gets me focused on ways to generate an impact.

About thirty people attend my session, which is part of a

breakout panel on "Entanglement and Marine Debris." They pepper me with questions afterward, seeming keenly and genuinely interested. They want to know what can be done, and they throng the "evidence table" I've laid out, loaded with items culled from the gyre: chewed-up soap bottles, umbrella handles, disposable lighters, toothbrushes, plastic bottle caps, and Baggies full of fragments. Judging by the reaction, plastic flotsam, the real deal, is worth about a hundred pictures, probably a thousand words.

I'm glad to spot in the crowd the weathered squint of Bill Macdonald, proprietor of Macdonald Productions, based in Venice Beach, a little north of my home turf. I'd run into him earlier in the hotel lobby, updated him about my new mission, and invited him to my talk. I first met him in the mid-1990s when I purchased one of his Hi8 underwater camera systems to document *Alguita*'s early adventures. (Unfortunately, its tenure with us was brief: *Alguita*'s early mishap on the Central Queensland coast inflicted fatal water damage.) I was impressed by his work with the Cousteau Society, both as videographer and as "marine awareness coordinator." He's given something like 350 Cousteau Society presentations around the country. In fact, I invited him to crew on one of *Alguita*'s early cruises, but he was tied up with shark work for the Discovery Channel. His is a hardheaded brand of environmentalism I strongly admire.

After the talk, Macdonald bides his time at the periphery. As the crowd thins, he moves in, commends the presentation, and is gone. But it won't be the last I see of him. On day two of the conference, he tracks me down. He tells me he hasn't slept well, thinking about my talk and letting the implications sink in. As someone immersed in the world of marine conservation, he says he's bothered that he'd lacked awareness of this "immense issue."

I know the feeling. He's says he's tired of shooting footage for the "Shark Week" crowd and wants to do something more meaningful. Then he makes his move. He suggests we coproduce a video about the "plastic plague." He quickly convinces me that a film will spread the word with far greater impact than the occasional PowerPoint. And he's already come up with a title: *Synthetic Sea*. I'm sold.

Macdonald wastes no time. He shows up at my house soon thereafter, loaded with cameras and brimming with ideas. I easily got the Algalita directors on board, but there's no room in the budget for a film on top of our restoration work and day-to-day expenses. But we know our issue is nothing if not visual, and a film could really bring the message home. I decide to fund the film myself, with the far-fetched hope of recouping my investment through sales of the finished product. With my own funds at stake, I'm thinking like a Hollywood producer. We've got adventure, sailing the high seas in a fifty-foot catamaran. We've got gonzo science. We've got a salty protagonist on a quest to save the oceans. But above all, we've got a message. Remember, in the year 2000, the Garbage Patch was still a well-kept secret. But I'm a performance novice and wonder if I can pull this off. My nerves settle down when I realize it's not about me; it's about the plastics plague.

Aside from the fish, there's a lot going on. We're waiting for the rainstorm we need to complete a new study of plankton versus plastic, this time in coastal waters, and other projects are brewing. We can't drop everything and head for the gyre, at minimum a three-week trip, to shoot this film. We decide instead to run out to a calm area on the east end of Santa Catalina Island, the low-budget Hollywood stand-in for a generic tropical isle. Its clear blue waters will ably double for the gyre. But we'll shoot early scenes on land, mainly on the banks of the Southland's concrete rivers, which often appear to flow with more trash than water. Macdonald and I work up a script and soon he's interviewing me on tape, sitting on *Alguita*'s deck. I'm holding forth about the emergence of plastic trash as a marine scourge, thinking it's going pretty well, until he gently informs me I'm looking everywhere but into the camera. He tells me to relax and just be myself. I wonder who else I can be, but nonetheless it's somehow reassuring. Thus, I begin to get my media legs. I don't know how I'd have felt at the time if I'd known my future held network news segments, late-night talk shows, a raft of documentaries, and countless YouTube postings. I get calls from old classmates who've caught me on TV and ask if I'm the same Charlie Moore they used to know—a legitimate question, since 4,869 Charles Moores currently reside in the United States. It may be a good thing I attained this notoriety in the wake of my first gray hairs.

I guide Macdonald to sites along the river and coast that I know to be choked with plastic debris, such as the San Gabriel River close to its ocean terminus, and Ballona Creek near Los Angeles International Airport. His camera captures heart-rending images of sandpipers and gulls foraging for food in a plastic-clad habitat. He works solo, using no assistant, giving him unfettered flexibility. And when the rains finally come, he's back—Johnny on the spot to get the "first flush" of floating trash that backs up behind the debris boom across Ballona Creek. It's a mother lode of buoyant garbage direct from the heart of L.A.

We visit the Sea Lab, our new research center in Redondo Beach, where Macdonald gets shots of me sorting through gyre samples. I serendipitously find a piece of orange plastic that's uncannily similar in shape and size to an edible amphipod, both on the same petri dish. This won't be the only time we highlight plastic's mimicry of edibles in the marine environment. On a later voyage through the gyre, we conduct an impromptu experiment. We catch a live salp colony and place it in a small glass fish tank sprinkled with trawled fragments. Then we watch in amazement. With video rolling, our little filter-feeding subjects pump their way around the tank and, sure enough, gulp plastic bits as if they were delectable plankton morsels. We later incorporate this scene into our presentation material.

I suppose we should have been prepared for the attacks by the plastics industry. A few years into my campaign, when we were getting real traction, plastics industry "spokespeople" attempted to discredit the salp clip by accusing us of creating a "staged hypothetical." Indeed, we soon realized they were attempting to discredit all of Algalita's research. Among the arrows in their quiver is the well-honed, battle-proved tactic of labeling unfavorable non-industry research as "anecdotal science." Sometimes the term they use is *junk science*. I silence them by serving up unassailable documentation of plastic-crusted salps in situ, including specimens we snagged on the first research voyage. I also point out that this is how science works. Indeed, even plastics were developed by people who staged experiments. It's a common and creditable scientific practice.

We hear from *Marine Pollution Bulletin* and learn that the paper has survived peer reviewing, but we're asked to clarify a few points before the article goes to press. The upcoming study of plankton and plastic in coastal waters moves forward. Seasoned scientist that he is, Steve Weisberg wants the new research to be accomplished before the first is published. He's still concerned that the first study will be open to criticism and possibly discounted for having been carried out in an area of low biological activity. The new study will tell us whether the plastic-to-plankton ratio is similarly troubling in nutrient-rich coastal waters that teem with biological activity. We'd expect more plankton in shore waters, but we don't know how extra debris from the L.A. metropolitan area will behave in choppier coastal waters and breezier conditions. It could be more vertically dispersed throughout the water column and inaccessible to our trawl nets. The prospect of comparing mid-gyre plastic flotsam with whatever's lurking in coastal waters may not sound terribly exciting to most people, but it is to us. We also can't deny the element of risk in this mission. If we can't demonstrate that plastic particles rival plankton in coastal waters, our cause may be weakened, although not in our eyes. Contamination of the gyre is cause enough for us. But if we've learned anything thus far, it's to respect the countervailing forces of doubt.

So many questions. Which of the two ecosystems will have the bigger pieces, the most broken-down bits, the most weathered fragments? How will the resin types differ? What will be growing on them? We know the central gyre vortex pulls in and traps debris from throughout the North Pacific, including tremendous input from densely populous Asia and the fishing fleet. But the vortex center is the doldrums, where plastics bob on placid surface waters. By comparison, near-shore waters are in a constant state of agitation. Even buoyant plastics can be dragged down by circulating sand and sediments. We also know the strong California Current sweeps debris southward, and prevailing westerly winds and surf may toss urban trash back onshore. Still, it would be surprising *not* to discover a serious plastic presence in these waters just west of a vast kingdom of fast-food joints, mini-marts, outdoor amusements, sports venues, and beaches visited 17 million times each year by sun worshippers equipped with food, drink, lotions, shovels, pails, Frisbees, beach balls, skimmers, and foam coolers—all plastic or brought in plastic.

Again with help from Steve Weisberg and Shelly Moore of SCCWRP, we map two sampling runs, each with trawls taken close to shore and farther out. We schedule the first run for October 2000 after sixty-three days of dry weather. The second will have to wait till after the next substantial rainstorm, putting us at the whim of one of the world's least rainy urban areas. Our reasoning is that storm runoff will push a big load of fresh plastic trash into coastal waters. The objective is to get snapshots of both sets of conditions, to correlate them, and to average them. The result should represent reality.

IT'S EARLY SUMMER. The marine air has a chilly bite as we pull out of the slip at first light. Leaving the California coast behind, the catamaran skims over choppy seas that make for a mildly bumpy, splashy ride. Macdonald, a few others, and I are finally headed for Catalina. He is a seasoned mariner as well as expert diver, more widely traveled than I in the South Pacific, Caribbean, Atlantic, and Mediterranean on Cousteau's famous research vessel, the converted British minesweeper *Calypso*. No stranger to the life aquatic, he helps crew on the twenty-six-mile trip out to Santa Catalina Island while briefing us on the filming plan. We've brought the manta trawl, and once in the shelter of Catalina, where the sea is glassy calm, we set the stage. The manta trawl and Macdonald, swathed in neoprene, go overboard. He has developed a method of balancing his underwater video camera on a boogie board, giving him a sea-surface perspective, a plankton's-eye view of the world. The trawl is meant to be a reenactment. We've brought debris samples from the gyre voyage to add verisimilitude. But the reenactment trawls actually scoop up plastic bits that we had no idea were there. Macdonald paddles in close to get a dramatic shot of the manta swooshing by and later admits he was worried about getting tangled in the trawl line.

Macdonald proves himself the Clint Eastwood of marine documentary filmmaking: highly efficient, few takes, knows what he wants. In the editing room, he draws on his extensive video archive to splice in compatible footage, and he wheedles additional tape from park rangers at Midway Island showing the tragic results of Laysan albatross chicks fed ocean-caught plastics by their well-meaning parents. I'm called up to Venice to record voiceovers, as the microphone picked up every little whoosh of wind, splash of water, and throat clearing. Macdonald says to start planning the premiere. He'll have a finished product in a matter of weeks.

WE DO OUR coastal study trawls and find as we analyze the results that our hypothesis is amply supported, with a few twists that cause us mixed feelings. Our data sets show that plastic abundance in the mid-Pacific gyre (that is, individual pieces per square kilometer) is a mere third of what's floating in coastal waters. Yet the density of gyre debris (the combined *weight* of all the individual pieces in a given area) is dramatically greater, by a factor of 17. Here's what seems to be happening: we assume the fragments in coastal waters are newer to the marine ecosystem, so it could be they've thus far eluded local filter feeders. Litter sequestered in the gyre has greater exposure over time to potential "predators," such as filter feeders that would vacuum up the smallest fragments. We're also aware from our deeper gyre trawls that a fair number of algae-fouled filament scraps sink and don't get figured into with our surface trawls. Counterintuitively, in urban coastal waters we find a superabundance of plastic fragments—up to eight million per square kilometer. The gyre study yielded an average 334,000 per square kilometer, which was jaw-dropping at the time but now seems pretty modest by comparison. Not surprisingly, the plastic count soars in our post-rainstorm trawls, which finally happen in January 2001. In certain transects, the plastics-to-plankton ratio far exceeds the gyre's 6-to-1 ratio. We find that before the storm, the plastics-to-plankton ratio is highest near the shoreline, tapering off farther out as the shadow of civilization recedes. After the storm it's a different story. Storm runoff pushes plastics farther out to sea, where they then greatly outweigh plankton.

Our mixed feelings stem from being right. Finding new areas of plastic pollution is not cause for joy.

What does it mean? As with much of science, our findings represent not an end but another step forward, with many more still to take. We're quantifying the presence and hope to learn the impact of plastic trash in the oceans. We're establishing baseline data against which future measurements can be compared. This is the only way we'll know whether plastic pollution is getting worse or lessening. This is how we'll learn if measures to reduce plastics pollution are working. We're on two tracks: science and reform. Armed with new information and firm convictions, we're Paul Reveres with PowerPoints. But there's still much to learn. Is mid-gyre plastic mostly from commercial and fishing vessels, or has it escaped from land? It's clearly a mix of both, but enforcement efforts need to target the worst offenders.

And there's still that matter of harm. What exactly is our plastic waste *doing* to the marine ecosystem.

We've staked out microplastics as our cause, so our line of inquiry will need to shift to ingestion. The oceans' primitive small fry, the tiny creatures that filter feed in surface waters, have evolved with limited discriminatory mechanisms. Once upon a time, virtually everything out there was good, digestible food. And it was one that these little guys provided healthy meals for food-chain higher-ups—not only for krill and small feeder fish but those great baleen whales. But we've yet to study ingestion per se, and thus can only claim that "potential" for ingestion is strong. We've also seen figures that suggest at least half of plastic flotsam gets weighed down by algae and other opportunistic organisms and sinks to the benthos—the seabed. There they join dense plastics, such as hard polystyrene CD cases and ballpoint pens, and PVC plastic things, which sink as a matter of course and could be harmful to the benthos. Anthony Andrady and others have proposed a "yo-yo" theory that works like this: algae and diatoms (phytoplankton) attach to plastic debris and proliferate in a process called "fouling." Now dense with marine vegetation, the debris begins to sink until it descends below the photic zone— beyond the reach of the sun's rays, which these little plants require to do photosynthesis. As the light-starved phytoplankton withers, mid-water bacteria go to work to digest it. Now cleaner and lighter, the debris piece resurfaces, ready to restart the cycle.

We may still be in the early stages of knowing the precise impacts of plastic litter. But we're no longer—if we ever were—in the early stages of thinking plastics have any rightful place in the marine environment.

From the beginning, there's been a faint aroma of something

sinister about the droves of plastics, the millions of tons of them, swimming alongside living organisms in the oceans, aside from the well-studied and deplorable issues of entanglement and ingestion. But even as late as 2000, after two gyre voyages and reading a good deal of plastics research, my assumption, shared by most of humanity, is that plastic, the material, is fundamentally inert. After all, this amazingly versatile man-made material has conquered all others and infiltrated every corner of our lives. *Inert* was always the operative word for plastics. Isn't that why things like baby bottles, milk jugs, and disposable foam coffee cups are made out of plastic? Cheap, strong, lightweight, and, of course, safe. By the 1980s, almost everything "baby" is made of plastic: strollers, cribs, car seats, mattresses, mattress covers, toys, portable tubs, teething rings, squeaky bath books and rubber duckies, brightly colored dishes, bottles, and sippy cups. Baby can fling her bottle or plate to the ground and it just bounces! Baby powders, shampoos, oils, diaper wipes, children's Tylenol syrup . . . all in plastic containers. Disposable diapers, nearly 100 percent plastic, are now the single most abundant garbage item in not-so-sanitary landfills. So plastics must be super-safe! If not, all hell would break loose, with whistles blowing, hearings, bans, lawsuits.

Or maybe not.

For decades, plastics got a pass. The few dissenting voices that questioned the safety of plastics were ignored, muted, or marginalized. Later research opened my eyes to the fact that early skeptics were speaking out almost from the beginning, in the 1950s and sixties. Learning this fortified my growing conviction—now speaking as an activist—that unpleasant truths will not be acted on, let alone listened to, until the major moving parts in a culture are aligned for a paradigm shift.

The first hint of trouble emerged in the 1990s, when suddenly we heard warnings about microwaving food in plastics. So maybe in that one situation, involving heat generated by molecular agitation, there was an issue with plastics. If I'd paused to think about it, I, a chemistry major, could have realized that, say, Saran Wrap exposed to microwaves could possibly spike those warmed-up leftovers with vinyl chloride—because, at the time, that's what Saran Wrap was, polyvinyl chloride, PVC film. But even if I'd had this clever perception, I wouldn't have known Saran Wrap also contained an additive, a phthalate, that made it so conveniently stretchy, and this might have imparted a nice dash of endocrine-disrupting compounds to my leftovers. We had yet to learn that these phthalates, found in personal care products as well as plastics, have a feminizing effect on developing males. I also recall reading that that "new-car smell" wasn't good for you. So you rolled down the windows, a simple enough remedy, without necessarily pondering the precise nature of the health risk posed by chemical off-gassing. And who hasn't had the experience of taking a swig of water from a plastic bottle left in a hot car and thinking, *Yech!* My partner Samala loves new-magazine smell, and despite my warnings can't resist deeply inhaling the aromatic and potentially neurotoxic fumes of the latest issue.

On the other hand, industrial toxics had long been on my personal radar screen. In my furniture shop from the 1970s thru the mid-1990s, I was worried all the time about my staff's exposure to volatile organic chemicals: the fumy solvents, lacquers, sealants, and finishes routinely used to strip and restore furniture—old-school paint strippers like methylene chloride, which evaporates in a flash and puts almost every biological system at risk of damage. In hindsight, though we took normal precautions, I realize we should have made better use of our respirators, especially when repairing vinyl Naugahyde furniture in homes and restaurants. This required mixing vinyl ingredients and heat-curing them on-site in an invisible cloud of fumes.

As plastics began sneaking into our lives—first a trickle, soon a flood, always looking like the new, improved version of whatever they replaced—acceptance was reflexive, and innocence was presumed. What a difference a decade or two can make. "Then" seems a time of magical, uncritical thinking, of being blissfully blind to myriad clues, and of misplaced trust in industry and government's benevolent oversight of the products in our lives.

That plastics are hydrocarbons, derived from oil, means they harbor the potential for toxicity, because oil, as we know, is inherently toxic. But there was something else going on with them. The first hint of that came as I thumbed through articles in the marine science library. I came across a study conducted by a South African wildlife biologist named Peter Ryan. Starting in the 1980s, he made it his business to study plastic ingestion in seabirds in the Southern Hemisphere. I was surprised to learn that seabirds—a distinct class of avian whose meals are exclusively taken at sea, at least before coastal garbage dumps—have been considered barometers of plastics pollution since plastics started polluting, basically in the 1960s. Soon after, Dutch scientist Jan Andries van Franeker began dissecting stranded (in this case, dead) northern fulmars collected from North Sea beaches, as a way of measuring trends in plastics pollution, which he initially found to be increasing. Ryan was able to correlate plastic ingestion in the birds he studied with detectable levels of toxic legacy chemicals (POPs) in their tissue and eggs. But he could not definitively prove that the chemical pollutants, in this case PCBs, came from the plastics and not from contaminated natural foods or other types of exposures.

The primary POPs at this time were eggshell-thinning DDT, the pesticide, banned in the United States in 1972, and carcinogenic PCBs, industrial lubricants, fire retardants, and coolants, banned in 1979. But, being persistent, they wouldn't go away. These synthetic molecules proved not only nearly indestructible but highly peripatetic. They wound up everywhere, including the oceans. But the key concept was this: it's in the nature of these chemicals, given their oil-based essence, to be drawn to like substances—fats, oils, lipids. All living creatures are composed of three basic ingredients: carbohydrates, proteins, and lipids. Thus, we attract and indeed harbor POPs, as do the creatures of the contaminated oceans, and as do, I was realizing, the processed hydrocarbons we know as plastics.

It was Shelly Moore, the SCCWRP biologist and statistician, who showed me the groundbreaking study by five Japanese researchers at the Tokyo University of Agriculture and Technology. She came across this paper in the American Chemical Society journal *Environmental Science & Technology* in early 2001 and immediately sent it my way. Titled "Plastic Resin Pellets as a Transport Medium for Toxic Chemicals in the Marine Environment," it contained proof, not mere evidence, that plastic floaters in near-shore waters were adsorbing (being coated by) wayward toxic chemicals. And in a big way. In their multifaceted and cleverly designed study, the researchers focused on polypropylene (PP) nurdles, the preproduction pellets that are raw material for most plastic products. As such, they're a major international commodity, shipped everywhere. PP—aka #5—is a strong plastic used to make bottle caps, food and sauce containers, stain-resistant carpeting, floating ropes, all-weather gear, and the like. Using sterilized stainless-steel tweezers, the research team collected nurdles from both industrial shorelines and recreational beaches in Japan. These were "feral" nurdles, coming from shipping spillage and escaped from processing facilities, inhabitants of the marine environment before washing ashore. A separate nurdle group was populated by "virgin" pellets secured from Grand Plastics in Japan. These were divided among several submersible baskets and secured to a wharf in industrial—that is, polluted—Tokyo Harbor. Each week a basket was removed until all had been collected, so that direction of exposure and rates of contamination could be gauged.

Their findings shook our world. In their lab, the pollutants were extracted from the pellets with hexane, a powerful solvent, and measured using advanced instruments. The basic results were this: the longer the nurdles remained in the pollutant bath—Tokyo Harbor—the more contaminated they became. They were pollution sponges. But even the most contaminated of the once-virgin nurdles weren't nearly as toxic as the shore-harvested nurdles. Those taken from industrial shorelines had toxic loads a million times stronger than toxic levels in nearby coastal waters. Nurdles from the cleaner beach sites were less contaminated but still dangerously polluted.

The ramifications were enormous. The special problem with nurdles is their strong resemblance to fish eggs—seabird caviar. Indeed, wildlife biologists like Peter Ryan had already established that preproduction plastic pellets were virtual staples in the diets of several seabird populations. Studies even managed to correlate seabird health problems—low leukocyte levels indicating weakened immunity—with heavier plastic ingestion, though without being able to prove causality, only a suspicious-looking correlation. Some seabird species retain non-digestible matter in their proventriculus—a pocket in the stomach holding indigestible material—for up to seven months. A toxified nurdle hanging around for seven months in a seabird's gut would not be a good thing. The Japanese researchers made another disturbing discovery: as the nurdles decayed, they released a toxic chemical identified as nonylphenol, a common additive that slows oxidation and, ironically, decay. It has also been proved a potent disruptor of cell behavior in lab experiments.

Since seabird biologists view their subjects as barometers of plastic pollution, the Japanese group saw plastic pellets as potential tools for assessing marine toxicity. How perversely fitting: a pollutant monitoring pollutants.

I remember the study Kathy Cousins told us about at the Derelict Fishing Gear conference in Honolulu. She'd seen data that appeared to correlate thin eggshells and hatching failure with plastic litter near Laysan albatross nests. I try to find the study but can't, and conclude it remains unpublished. This eggshell thinning is a page out of Rachel Carson's *Silent Spring*, one of the damning effects of exposure to DDT. Carson died in 1962, more than three decades before researchers discovered the insidious propensity of oil derivatives to alter hormonal signaling and even gene expression in biological systems, including humans.

With my mental neon still flashing *Show harm!* I know the Japanese study belongs in my maiden scientific paper, not yet in press. We send an e-mail to the *Marine Pollution Bulletin*'s editor, Charles Sheppard, asking if it's allowable at this stage to add a line and a citation. He's agreeable. In the first paragraph, we insert these words: "In addition, a recent study has determined that plastic resin pellets accumulate toxic chemicals, such as PCBs, DDE (a derivative of DDT), and nonylphenols, and may serve as a transport medium and source of toxins to marine organisms that ingest them."

And with this comes a far greater sense of urgency. Indeed,

I'm surprised by how far I got, starting in 1997, motivated solely by a simple yet intense annoyance about a litter problem in the mid-ocean. And while it seemed a given that planktonic and filter-feeding marine organisms are eating the little plastic bits that have so captured my interest, now I wonder if they're also being poisoned. Do microplastics kill these little creatures? Are they a threat to the entire food web? Initially I saw the trashing of the ocean's beautiful, faraway waters as a symptom of man's disrespect for the natural world. But this toxic element represents more potential harm that even I'd bargained for.

SYNTHETIC SEA STARTS with images of dolphins torpedoing ahead of a sailboat and Bill Macdonald's solemn voiceover intro. The camera cuts to me in my yellow oilskin, standing by the winch table, hauling in the manta trawl. I say: "As captain of the oceanographic research vessel *Alguita*, I've traveled to many remote areas of the Pacific Ocean. And in my travels I've been alarmed at the increase in the amount of trash—plastic debris—on all the beaches that I visit. My sentiment was that the ocean is filling up with trash."

Nine minutes later, it's over. My expectations are exceeded. In these few minutes, the entire issue is laid out concisely, vividly, and scarily—including the toxic-nurdle study that emerged from Japan as we filmed. We've designed the film to be a "short," something to be embedded in longer presentations we'll give at college campuses and environmental gatherings. *Synthetic Sea* ushers in a new era, not only for me and the foundation, but for the campaign against plastic waste and pollution. We honor Macdonald by offering him a seat on the Algalita board, a role he fills ably for five years. And with *Synthetic Sea* he's only beginning his career as chief documentarian of the plastic plague.

A public premiere—shortly after 9/11, as it turns out—is hosted by Eco-Link, a local environmental group that is also honoring Captain Paul Watson for his efforts to stop the U.S. Navy's testing of underwater sonar detection systems, known to injure and disorient whales and dolphins. In another six years, Watson will shoot to TV fame—or infamy, depending on your point of view. As a modern day eco-Ahab on *Whale Wars*, he seeks out and harasses Japanese "research" whalers in Antarctic waters. After the event, we spend time with the very intense Paul and his soon-to-be third ex-wife. His views are fascinating and discomfiting. Collapse of the marine ecosystem is inevitable, he says, though which part will go first, and when, he does not claim to know. He tells us *Synthetic Sea* only served to reinforce this belief. His captaining of Greenpeace ships and his own Sea Shepherd Foundation vessels have taken him across every ocean, and he, too, has noticed the growing intrusion of plastic trash. He likes to stop the boat each day for a swim, he says, and there's always trash in the water. As much as Watson loves the ocean (as a teen he rode the rails from his landlocked Canadian hometown to join Canada's coast guard) he's most ardent about animal rights, especially those of marine mammals. He's laid his life on the line for them, both in the Antarctic, as seen on TV, and in eastern Canada, where he's physically shielded snowy-white Canadian harp seal pups from hunters' clubs. Many of us strongly believe animals deserve to live free of man's predations and habitat-fouling ways, but few of us put our beliefs to work the way Watson does.

Ten years on, I ask Bill Macdonald to remember the reaction to that Eco-Link showing of *Synthetic Sea*. He says, "It was overwhelming. Better than expected. The issue could now be debated, since all the 'damning' information was from published scientific papers." And I have, by the way, made back that investment.

DECEMBER 2001. A year and a half after our 1999 research voyage to the gyre, and three and a half years after the fateful first passage through the gyre, we receive a copy of volume 42, number 12, of the *Marine Pollution Bulletin*. Here it is: "A Comparison of Plastic and Plankton in the North Pacific Central Gyre." This straightforward five-page study—much of it taken up by a table, two graphs, and a chart—has proved a solid workhorse. The average study might receive a few dozen citations in subsequent research papers. So far this one has chalked up over eighty (and counting) citations in published and peer-reviewed papers. I take from this two things: first, that plastics pollution in the marine environment has become a hot topic, the kind graduate students choose for thesis research. And, second, that even an unaffiliated, uncredentialed, independent scientist such as myself can get a fair hearing with the right investigation—though it never hurts to get a little help from well-credentialed friends as coauthors.

Between the paper and the video, my dance card seems to be filling up. Speaking invitations multiply and now we have a strong little film to show. We lay plans for another voyage into the gyre, and a trip to a tropical shore claimed and defiled by plastic debris.

CHAPTER 11

NET LOSSES

SUMMER 2002. It's not like booking a trip to Maui. The permit process is torturous and the fees stiff for gaining access to the ecologically fragile Northwest Hawaiian Islands. But now *Alguita* lies at anchor in an atoll's sheltering, shimmering lagoon, encircled by 20 miles of crescent reef and sand bars. Well, not so sheltering. We've had to lash the catamaran's stern line to a decrepit metal seawall, installed during World War II to buttress a man-made runway. The currents here are unpredictable and strong, capable of pushing us back into the fringing reef. We listen for the small plane that will bring Curtis Ebbesmeyer and Jim Ingraham to this place, called French Frigate Shoals, 570 miles northwest of Honolulu. These affable oceanographers are technically retired, but carry on as if not.

This is an interesting place, French Frigate Shoals, more like a shallow thirty-mile-wide crater than anything resembling a postcard atoll with swaying palms and golden sands. The total land area—spread among twelve sandbars—is a mere sixty-four acres, but the lagoon is the largest in the chain at two hundred square miles. Darwin himself was first to describe how atolls come to be: a volcanic island gradually sinks beneath the sea surface while surrounding reefs keep growing upward. Corals stop growing at the so-called Darwin point, where ocean waters become too cool to support them. The atoll's odd name honors an eighteenth-century French explorer, Jean-François de Galaup, comte de La Pérouse, whose two frigates brushed up against the atoll's hidden reefs—marine equivalents of land mines—one dark night in 1786. Given our struggles with these currents and twenty-knot gusts, I'm not surprised. It doesn't help that electronic charts of the area put our position on land when we're at anchor.

The largest sandbar, Tern Island, is essentially man-made, a spit enlarged by dredge and fill during World War II. By design, it resembles the deck of an aircraft carrier. It accommodates one of just two working airstrips in the Northwestern Hawaiian Islands; the other's on Midway, a five-hundred-mile hop northwest of here. The strip runs along the twenty-six-acre island's central ridge, elevation six feet. Near the airstrip is a weather-worn former Navy/Coast Guard station that houses two year-round U.S. Fish and Wildlife Service fieldworkers, who watch over the atoll's rare, protected fauna. The station will serve as lodging for the oceanographers for two nights. A catchment system collects rainwater. Vegetation is sparse. This is an aquatic rather than a terrestrial habitat, with life teeming below the sea surface.

But why are we here? One reason is that *Alguita* has been chartered as a platform for spinner dolphin research by Texas A&M scientist Leszek Karczmarski. We'll make our base here for a week while he observes a local dolphin population and takes DNA samples by grazing the leaping spinners with a strip of Velcro attached to a long pole. Another reason lies in a utility shed near the airstrip, evidence of one of nature's crueler ironies. These remote islands are federally protected marine wildlife sanctuaries, but they are also world-class flotsam repositories. This makes them unintentional natural laboratories for the study of marine debris, mostly plastic, and the havoc it wreaks.

In the utility shed is stored stranded trash gathered on Tern Island by staff over the last 112 days. Ordinarily they walk the perimeter every two weeks, documenting their finds and saving them for periodic removal. I arranged for this special collection with Fish and Wildlife Service brass, who are also keen to know more about the flotsam collecting on the atoll's shores and reefs. Once the plane arrives, Ebbesmeyer will begin going over debris I've already categorized and arranged on makeshift tables and mats. Ebbesmeyer's role is to quantify the contents and take a stab at identifying its origins. His quarry won't include net clumps hauled by the tens of tons from the shores and waters of the Northwestern Hawaiian Islands each year. Nor will it include microplastics I trawl from waters in and around the atoll. There's so much big plastic here that the small stuff barely registers as an issue. Nonetheless, it's here, and in the lee of French Frigate Shoals I learn an important lesson: sea state correlates strongly with trawl "productivity." In a roiling sea, the smaller plastic bits whirl downward and deeper, beyond the reach of the manta net. In a calm sea, they drift back up to the top. It's logical, of course, but it's also a confounding factor requiring verification. We trawl in choppy seas and get very little. When the trades ease, we get a tremendous amount, in the same general area on the leeward side of the lagoon.

French Frigate Shoals is one of the more southerly North-western Hawaiian Islands, a low-slung 1,200-mile strand of 10 named islets, atolls, jutting rocks, reefs, and shoals. Here you'll find no hula revues, tiki bars, or souvenir stands. Casual visitors aren't welcome, and vetted seasonal guests are mostly scientists and eco-volunteers. It's for good reason. For millennia these islands were protected by their natural isolation, making them safe havens for large endemic populations of seabirds, sea turtles, spiny lobsters, tropical seals, and a few thousand other species. Then came "discovery." The nineteenth century brought men with clubs from Japan and America to "harvest" nesting birds— with no evolutionary impulse to flee—for feathers, and basking seals for their pelts and oil.

The Northwestern Hawaiian Islands were once proper, robust islands, and they foretell the fate of their younger southerly neighbors, the "main" Hawaiian Islands, in a few tens of millions of years. Out here in the middle of the largest, deepest ocean, rooted in the earth's shifting crust, the atolls slowly erode and subside, eaten away by waves and drawn downward by the cooling and contraction of their volcanic bases. Climate change and rising sea levels could mean a quicker end, and serious habitat loss for threatened species. The official U.S. government website for the Northwestern Hawaiian Islands admits the islands are "quietly slipping into the sea," on their way to joining other, older fully submerged seamounts called guyots. If you check NASA's satellite map for the Hawaiian Islands and zoom in on the thread of tiny aqua specks northwest of Kauai, you get a sense of their heartrending vulnerability.

But strict laws meant to protect the Northwest Islands from predation and exploitation are meaningless to the fifty to sixty tons of plastic debris that strand here each year. Most of it is fishing net and gear, but there's plenty more of unweighed miscellaneous plastic garbage of the type I've sorted in the shed near the runway. It drapes the reefs, wafts in the lagoons, and heaps on the quiet shores, more than two thousand miles distant from the closest major population center. Oceanography explains it. The Hawaiian archipelago bisects the North Pacific's east–west currents, like stepping-stones across an easterly flowing stream. Its northern tip is the closest bit of land to the debris-rich Convergence Zone, the area that connects and overlaps Ebbesmeyer's eastern and western "garbage patches." When garbage-laden currents rake through the chain, flotsam gets strained out. We say the atolls and islets are like teeth of a comb pulling debris from the ocean.

But the problem isn't the currents. It's what they contain. Fifty years ago, these shores would have sifted out driftwood, glass floats and bottles, and scraps of hemp net and rope, not man-made plastics by the ton, much of it lost, abandoned, or discarded by the fishing fleet. Research shows that the amount of debris deposited each year varies with natural cycles. During El Niño years, when equatorial waters heat up, the Convergence Zone shifts southward, closer to the Northwestern Hawaiian Islands. In these years the deposition rate soars, creating special peril for one of the world's most endangered marine mammals, the Hawaiian monk seal, an animal that's especially prone to entanglement.

It happens that French Frigate Shoals is home to the largest of six Hawaiian monk seal colonies in the Northwestern Hawaiian Islands. This is the largest, though not the healthiest, colony in the world, with close to four hundred members. Here in the lagoon we catch glimpses of them. They circle the reef, hunting for food, and haul out on the sandbars to rest. You could say the Hawaiian monk seal is a superstar of endangerment. Despite strenuous decades-long efforts by wildlife and marine biologists, their numbers decrease by a relentless 4 percent per year. About

eleven hundred seals are accounted for, and only one in five new-
born seals survives the four to six years to sexual maturity. Unless the situation can be turned around, the natural population
could be extinct in a matter of decades.

I'm very impressed with these seals' curiosity. Food is sparse,
so they're obliged to investigate everything. One individual, a
heavy older animal sporting the characteristic monkish whiskers,
keeps nosing around the *Alguita*. Bothering or befriending monk
seals is verboten, and killing one, under state law, will get you a
fine up to $50,000 and five years in prison. Our would-be friend
might want a food handout, but it simply seems curious about
this novel feature in its habitat, a catamaran. I'm no marine mammal expert, but I can easily see how these engaging creatures
would be prone to entanglement because of this urge to investigate and to disturb things around them to find food. Stray nets
in the habitat would be like booby traps, enticing a seal with
bycatch food items.

For an up-to-date briefing on the seals, since its fortunes have
continued to decline since 2002, I check in with Bill Gilmartin,
a noted Hawaiian monk seal authority. He's a retired seal specialist for the National Marine Fisheries Service, a member of the
International Union for Conservation of Nature's seal specialist
group, and cofounder of the Hawai'i Wildlife Fund. He drafted
the first Hawaiian monk seal recovery plan in 1983 and remains
close to the front lines. Now he lives in Volcano Village, an
upslope artists' enclave on the Big Island, where he divides time
between conservation efforts and woodworking. And he'd be a
natural to portray John Muir.

Gilmartin suggests I have a look at the updated Hawaiian
monk seal recovery plan, approved in 2007. At 165 pages, it's
dauntingly comprehensive, yet makes for strangely compelling

reading. I learn that Congress generously funds efforts to save the seal, but, oddly, greater protection has done nothing to reverse continuous decline. Is the seal worth saving? Of course— not only for the sake of biodiversity, but to render justice. Absent man in its habitat, the seal would be flourishing, just as it was when early explorers brought back word of beaches blanketed by basking seals, and the pelts to prove it. The monk seal is special, being the only tropical pinniped and "a living fossil," resembling ancestors that swam the seas 15 million years ago. (Modern man appeared a mere 200,000 years ago.) Hawaiians called the seals "the dogs that run in the waves." Adult seals are big, averaging 400 to 600 pounds, and their natural life span is 35 years. A Caribbean cousin was hunted to extinction by the early 1950s, and a Mediterranean relative teeters on the brink. They spend two-thirds of their lives foraging at sea and dive to impressive depths seeking benthic (bottom-dwelling) prey. They "haul out" individually on isolated stretches of beach, unlike most pinniped species, which tend to congregate. Females sequester themselves when they give birth and nurse their pups the six weeks to weaning. Among reasons variously given for the name *monk seal* is this solitary urge, along with the whiskers and cowl-like folds of skin around their necks. Some accounts say the seals once populated the main Hawaiian Islands but retreated to the "suboptimal" but unpopulated northwestern archipelago after Polynesians arrived. The seal population at the time of Cook's discovery is anyone's guess, but one nineteenth-century ship returned to its New England port with 1,500 pelts—more dead animals in one haul than exist today on the entire planet.

Gilmartin begs to differ with the "suboptimal" contention. "The Northwest Islands have the reefs and lagoons, which have the food they like," he tells me—simply not enough of it. He also says they're clever. For example, they've been observed to flipper-lift stones in search of prey. "Crittercams" attached to seals reveal that opportunistic feeders like skipjacks will often shadow them and wait for a seal to flush out prey, then grab it away, which makes these fish pretty smart too. On Kure Atoll, the northernmost island in the group, young seals fed fish from a boat in early conservation efforts continue as adults to approach and peer into visiting vessels, evidently recalling the good old days and hoping for a free lunch.

The seal harvest ended in the early twentieth century, but the animals' fortunes failed to improve as military and commercial activities degraded their habitat. Soldiers hunted them for sport during the World War II military occupation of Midway and Laysan islands. With habitats now restored on Midway and Laysan, seals colonize six of the atolls.

In the post–World War II era the seal population climbed, peaking at about three thousand in 1958. Since then it's dwindled, stabilized, and dwindled again—this despite passage of the Marine Mammal Protection Act of 1972 and despite the seal earning official "endangered" status in 1976. Still, it wasn't until 1986 that the seal's habitat attained sanctuary protection. And yet commercial fishing continued, thanks to flawed and possibly biased academic science showing ample fish and crustacean stock for all. The seals had to compete with commercial fishermen, and they were getting snared in fishermen's longlines and nets. Reliable reports had fishermen clubbing, shooting, and poisoning hungry seals that tried to "share" catch that by rights should have been theirs. In 1991 the state of Hawaii banished longline fishermen from sanctuary waters, but lobster and "bottom" fishers remained, harvesting crustaceans and cephalopods (octopus and squid). With longline fishers gone, the seal population gained slightly, but only briefly, before ebbing again.

In 2000, the Earthjustice Legal Defense Fund, representing several environmental groups, successfully sued the National Marine Fisheries Service for dereliction of duty. The complaint accused the service of failing to discharge its duties under the Marine Mammal Protection and the Endangered Species acts by allowing lobster and bottom fishing in the sanctuary as the seal population declined. The attorneys marshaled new research that roundly disproved science paid for by fishing groups that claimed lobsters and cephalopods were not dietary staples for seals. In fact, seals prefer these bottom dwellers. The lawsuit led to permanent closure of the lobster fishery in 2001. But it was too late. The lobsters had been trapped down to near extinction and have yet to recover. The Earthjustice lawsuit is an example of a powerful tool a small group can wield, provided laws are on the books that require an agency to perform in a certain way and there's good evidence they're not.

Few seals die in plain sight. They vanish, and the yearly seal census records fewer, even with strong legal protection, geographic insularity, and well-meaning human allies. But Gilmartin says the remaining risks are hard to control and produce outsized effects on such a tenuously small population. Some are behavioral and include mothers who abandon their pups when disturbed or surprised, and male aggression toward females and pups, sometimes involving "mobbing" by groups. Once weaned, pups are abruptly on their own, competing for limited food with adults and other species, and highly prone to shark predation. Gilmartin also tells me the character of the aquatic environment has changed, with warmer waters resulting in "diminished productivity" of natural food sources for the seals. Field notes posted by wildlife managers are heartbreaking, often recording the sudden absence of a pup, or emaciation, which creates susceptibility to pathogenic disease.

In the recovery plan, risks to seals are enumerated and ranked, along with chances for mitigation. In all, eleven risks are cited, including the behavioral quirks. But the top three, considered "crucial," are food limitation, shark predation, and entanglement in marine debris. Talk about unintended consequences! As Einstein rued that relativity spawned nuclear weapons, polymer scientists should shudder to think how many animals have been killed by their miraculous material. My study of the issue convinces me that plastic debris is second only to commercial fishing as a killer of marine life and a greater immediate risk to the ocean's biota than a chaotically changed climate. We can limit our presence in the seals' habitat, but our wayward plastics drift into their lives, harming by proxy. Not "ours," really. Mostly those of the global fishing fleet that prowls the North Pacific, leaving in its wake "ghost nets" and monofilament longlines gleaming with hooks. Between 1982 and 2006, 268 entanglements of monk seals were documented. In 1999 alone, 28 seals were killed by derelict fishing gear. These figures are considered the proverbial tip of the iceberg. Many fatal seal-debris encounters are assumed to occur out of sight, beneath the waves.

The recovery plan notes that entanglements began to be observed in the 1960s, when "durable and resilient plastic materials . . . replaced natural fibers in the maritime industry." It also says that the 1989 implementation of MARPOL Annex V, prohibiting dumping of garbage including plastics in the marine environment, has had zero impact on the debris deposition and entanglement rates since 1983. The consequences are lamentable.

Some animals drag entangling nets around, becoming exhausted and unable to successfully forage food. Finally they starve. Others sustain deep gashing wounds that become badly infected. Others struggle and drown, or themselves become prey. And the animals most seriously at risk are the playful, curious young ones that Gilmartin says are pretty much just like frisky puppy dogs. Eighty percent of observed entanglements involve juvenile seals. The implications for species survival are gravely clear.

While the death toll by debris is unknowable, a December 2008 NOAA report affirms that "Hawaiian monk seals become entangled in fishing and other marine debris at rates higher than reported for other pinnipeds"—not because this seal asks for it, but because plastic debris engulfs its habitat. It seems a cruel twist of fate that this charming animal, termed a "charismatic megafauna" because of its appeal, finds itself in the thick of a synthetic blight. But it is only one among hundreds of species— and millions of marine animals—harmed or killed each year by plastic gear.

It's become very clear to me that any discussion about how the oceans came to be choked with plastic garbage needs to include the fishing industry, both its practices and the materials it uses. Derelict fishing gear includes not only lost and discarded nets and monofilament longlines but floats and buoys, traps, plastic barrels and crates, and an array of ordinary consumer items. Commercial fishing finds itself at the nexus of several marine crises. Lost and abandoned plastic fishing gear. The dumping of plastic garbage that includes instant coffee lids, butane lighters, and spent light sticks. Overfishing, enabled by the switch to lightweight plastic gear. And the "bycatch," or incidental killing of millions of marine animals each year, from whales on down. For example, a single Danish ghost net recovered in the Atlantic in the 1990s contained twenty thousand pounds of dead cod, a species now collapsed in the once fruitful Grand Banks off New-foundland.

In 1990, BirdLife International estimated that 17,500 Lay-san albatross had been killed by drift nets. This type of net was banned two years later, only to be replaced by longlines, which, according to the American Bird Conservancy, can be up to sixty miles long and strung with as many as thirty thousand hooks. The carnage increased, with annual albatross and petrel death tolls rising into the hundreds of thousands. The birds dive for the bait and get hooked and dragged to drowning deaths. New guide-lines generated by the National Marine Fisheries Service require American fishers to minimize longline bycatch by using rounded hooks or bait covers and attaching bright plastic streamers to their lines to frighten the birds. These methods have helped to a certain extent, and other countries have adopted them, but not all. Some fishing fleets are completely and unscrupulously rogue, making bycatch death a persistent and serious concern that con-tributes to "slow decline" among most seabird populations.

The International Whaling Commission considers "entan-glement in active fishing gear, derelict gear, and other types of marine debris," to be the primary cause of anthropogenic (man-caused) mortality in humpback whales and as many as 300,000 other cetaceans each year. They now consider bycatch a "pri-mary concern."

While researching this chapter, I decide to pay a visit to the International Bird Rescue and Research Center in nearby San Pedro. Here I learn firsthand the results of "fisheries interaction," as it's called. My guide, dedicated veterinarian Hayden Nevill, tells me, "Fishing-related injuries are mostly what we tend to see." She explains that wading birds, especially, tend to get in trouble with monofilament line wrapping around their legs and cutting through tendons and soft tissue. When this happens, the birds must be euthanized because they can no longer survive in the wild. She says a seabird oiled in a spill has a better chance of surviving than an entangled one. In a box on a counter is a rat's nest of tangled fishing gear, removed, she tells me, from a "couple dozen" birds. I see clumps of line to which are attached blinking and foam bobbers and homemade and artisanal lures used by sport, not commercial, fishermen. And it strikes me that the harm they did was preventable.

Plastics and commercial fishing have a seemingly perfect relationship. It's hard to imagine an industry better suited to make use of cheap, waterproof, lightweight, fantastically formable plastic. In commercial fishing, plastics proved revolutionary and they regrettably modified behavior. The new nets and lines made of nylon, polypropylene, and polyethylene filaments were a fraction of the weight and price of traditional gear. The weight and expense of organic materials used in traditional net making—fibers such as hemp, sisal, manila, and cotton—naturally capped net size and catch, and they were assiduously maintained and reused. Japanese blown-glass floats were replaced by hollow plastic balls and foamed plastics. The advent of cheap plastic gear ushered in an era of single-use practices, with many fishers calculating that fuel and the space for catch that was saved by jettisoning nets was worth it. Fishermen will even bundle and throw old gear overboard to attract tuna. Any alien object in the formerly debris-free ocean is of great interest to these inquisitive, intelligent creatures.

Freed from natural materials and equipped with high-tech tracking devices, the international fishing fleet is formidable. Three million vessels—from state-of-the-art factory ships to local scows—have depleted the ocean's larder by 80 percent. China's catch is four times that of distant competitors: Peru, the United States, Indonesia, Chile, and Norway. Until 1992, when they were finally banned, drift nets up to forty miles long voraciously emptied the oceans of marine life. Pieces of these "legacy" nets, and the yellow foamed plastic "banana" floats used to buoy them, are still among the most prolific debris items found, not only in the Hawaiian Islands, but along the Alaska shoreline. In that storage shed at French Frigate Shoals, Curt Ebbesmeyer counts more banana floats than any other item. He sardonically notes how odd it seems, since drift nets were banned ten years earlier, though smaller versions are still legal—even in Hawaiian waters, where I was shocked to learn one thousand near-shore gill nets are in regulated deployment at any given time.

The scourge of derelict plastic fishing gear has prompted a burst of international symposia and policy revision, including the Honolulu conference where I surreptitiously displayed our poster. A more recent Honolulu event was sponsored by Asia-Pacific Economic Cooperation. Called an "educational" gathering, it was convened by an international who's who of ocean policy experts. Its purpose was to helpfully expose high-ranking delegates from fishing associations in China, Japan, Korea, Samoa, and the United States to reports by leading scientists about the "adverse effects" of derelict gear. They learned a lot—for example, that an estimated million seabirds are killed each year by entanglement in longlines. And that 100,000 turtles and marine mammals—seals, dolphins, whales, otters—die annually, snared in their nets and lines. And that bycatch kills weaken their own industry, not only by depleting fish stocks, but by sabotaging the food chain. And that derelict fishing gear destructively coils around ships' propellers, including their own. And that debris-caused repairs, downtime, and fish stock loss are estimated to cost fishermen billions per year. And that even human loss of life can result from lost nets. The most horrific instance of this occurred in 1993 when a South Korean ferry caught in heavy seas capsized when a net clump seized its propeller. Two hundred and ninety-two passengers drowned. A study conducted in the aftermath found that "in a two-year period [1996–98] there were a total of 2,273 navigational incidents that involved vessels and marine debris in Korean waters, including 204 involving propeller damage, 111 involving operational delay, 15 involving engine trouble . . . and 22 involving 'disaster' [loss of vessel and/or people]." I cite this study because it offers a shocking glimpse of an international plague.

Efforts by people like me to bring about laws requiring the reporting of net losses as well as container spills have thus far failed. Fishing association representatives will tell you that stiffer laws won't guarantee compliance. Rogue fleets from poorly governed countries cannot be controlled. Among U.S. fishermen there are also rogues who aren't ashamed to put their outrageous behavior on display. A friend suggested I check out the Discovery Channel reality show *Deadliest Catch*, which follows the exploits of an Alaskan crab-fishing crew. The crew members shout, haul, and scramble around slippery decks; they swab the decks with gallons of bleach, then flick the empty plastic jugs into the ocean. They shoot an "unlucky" inflated plastic float with high-powered rifles, trying to sink it. The adjective that comes to mind is *flagrant*. I've seen my share of these blue jugs and plastic floats in deep ocean waters and caught countless blue plastic chips in my trawls. Under current U.S. Coast Guard regulations, this crabber's cargo should have been inventoried before and after the voyage to assure that all plastics and other non-perishables were accounted for. Obviously, enforcement, or lack thereof, is an apex issue. The oceans need to catch a break. Next time you're ordering dinner, consider saying no to fish and shrimp, and buy chicken-based cat food for your felines.

Cynthia Vanderlip is with us at French Frigate Shoals. She's the wildlife manager at outermost Kure Atoll, and she's been directly involved in seal recovery and derelict fishing gear removal. The debris gets tangled into Gorgon-like mixed heaps weighing upwards of a ton, and rigorously trained multiagency divers gingerly peel netting off reefs. Loose nets ruinously scour coral reef and benthic ecosystems, in addition to their other quirks. It takes rigorously trained divers to locate and gingerly remove them. Agency coordination has grown, which is a good thing. NOAA organized an effort to sensor-tag and track net clumps in the Convergence Zone, planning to pick them up later. We deployed seven of their satellite buoys on derelict gear on four different voyages. Data was retrieved, but as a removal strategy, it didn't work. The nets were too widely scattered, making costs prohibitive. A subsequent effort using a drone aircraft also failed. NOAA has now suspended its ghost net remote sensing and retrieval operations in the open sea. Kure's debris removal operation in late summer of 2009 was perilously carried out in high seas with swells up to eight feet. Four tons of derelict plastic nets were removed and loaded for transport back to Honolulu on the 225-foot Coast Guard cutter *Kukui*. During the weeklong operation, Vanderlip said, personnel rescued seven ensnared monk seals, five endangered black-footed albatross, and a tern. "The impact is devastating," she says. "It's such a serious problem and it can't be solved until the responsible parties are held accountable. But in the meantime, we'll continue to do what we can."

The fact is, it will take years to know if antipolluting laws and programs are making a dent. The reason is simple: the debris infestation has been accumulating for decades and has not stopped. In Hawaii, those gill nets made of plastic monofilament are getting loose more often, despite and seemingly because of tougher laws passed in 2008. Now fishermen who breach the rules are more inclined to abandon their nets and run when they spot state enforcement agents—yet another unintended result of well-meaning policy. No one, not even Ebbesmeyer and Ingraham, know what percentage of drifting debris winds up exiting the marine environment, either by stranding or being dragged to port by Good Samaritans. Neither do we have anything but estimates about the oceans' current plastic load, so there's no way to gauge the results of remediation efforts. We do know the land area available for stranding is minuscule relative to the ocean's size. Ebbesmeyer says debris in the gyre can be stuck there for fifty years, probably longer. But dealing with mid-ocean ghost net attacks is scary, dangerous, and expensive. Maritime insurance companies, not surprisingly, strongly support efforts to prevent fishing gear loss. They're out tens of millions each year paying damage claims.

In an attempt to nab the worst offenders, a team of scientists affiliated with NOAA and the University of Hawaii undertook a study of ghost nets recovered by NOAA and the Coast Guard in their annual Northwest Hawaiian Islands cleanups. They were able to document more than 250 distinctive net styles. A team of international experts was brought in to help identify the countries and fleets they belonged to. I contacted researcher Molly Timmers and NOAA's outreach officer, Carey Morishige, to see if the database had led to any apprehensions. Morishige reported that net identification programs are ongoing in Australia, where the northern coast is another debris magnet. Such efforts have also produced results in Puget Sound, where abandoned crab cages are a particular problem. But both affirmed that the program has been less successful in Hawaii. Timmers has since switched her research focus to coral reef habitats. Morishige explained in an e-mail:

> NOAA has since discontinued the collection of that type of data. . . . Essentially, the "configuration" and type of nets we get in Hawaii are large conglomerations of net pieces — not the entire net — which you would need for accurate characterization of type of net and possibly the fishery. Our results showed that most times you cannot find the source from these nets, perhaps the manufacturer, but not the fishery, location, country, boat, fisherman, etc., that lost or abandoned that net.

A trail gone cold.

For a recent update on the monk seals' plight from someone still in the field, I contact my crew member from 2002, Cynthia Vanderlip. I read in the recovery plan that the seal colony on Kure, the atoll she seasonally manages, was the only breeding site to buck the downhill trend. It is home to about one hundred seals. She famously photo-documented and freed a seal whose muzzle was clamped shut by a conical hagfish trap. Vanderlip thinks the injured animal could not have survived with her jaw muscles so atrophied. "We never saw her again," she told me. This was a young female that might have borne a dozen or more pups in her lifetime. As to the putatively healthy colony on Kure, Vanderlip said: "I think Kure is declining now. Some of the pups born this year were already looking thin when I left Kure in October. I don't think the ecosystems have recovered yet."

Back on French Frigate Shoals, toward the end of day three, Ebbesmeyer completes his tally: 199 litter pieces total had been collected on tiny Tern Island over 112 days. An independent videographer, Michael Bailey, has joined us. The plan is to get footage to add to *Synthetic Sea*. A major conference is coming up in the fall, set for Santa Barbara, and we want a bigger, better film to present. Ebbesmeyer is a natural performer. The cameraman enters his lair, the utility building, and captures the neatly sorted heaps of fishing gear and a few consumer containers as Ebbesmeyer describes the debris. Most plentiful are drift net floats, eighty-eight of them, both hollow plastic and foamed polyurethane. Once they were attached to the tops of miles-long nets hanging like lethal curtains in the ocean. Some of the floats are weathered, no doubt legacies. But others look suspiciously fresh, which they shouldn't, as drift nets have been banned for a decade. Second most common are plastic oyster spacer pipes, eighty-three to be exact. These are rigid plastic tubes used in Japanese aquaculture, more than two thousand miles distant. How did they get here? Ebbesmeyer says storms break up the coastal oyster farms and carry the spacers out to sea. Way out to sea. There are fewer than ten each of the remaining items, but most are also connected to fishing: light sticks, disposable lighters, cast floats, hagfish traps of the kind that muzzled the seal on Kure, and an unopened can of Miller beer. In all, 86 percent of the debris was fishing related.

Cut to the Tern Island shoreline. We're all a little taken aback when Ebbesmeyer declares for the camera that the human race may well be doomed by plastics. He gives us a few generations before hormonal toxicants in plastics render us chemical eunuchs. Maybe this is one of the seal's problems. But for now, no one knows.

CHAPTER 12

INDIGESTIBLE

The base of the marine food chain is being displaced by a non-digestible, non-nutritive component that is actually outweighing and in some cases outnumbering the natural food. That is our core issue.

—THE AUTHOR, at an Algalita board meeting

SIX YEARS AND a few adventures later, in January 2008, I find myself in Radio Bay, the port at Hilo town, one of Hawaii's lesser-known gems. With a fresh crew, two recruited from Hawaii, three flown in from the West Coast, *Alguita* heads out of port on its seventh foray into the gyre. This will be a new kind of fishing expedition—about fish as well as plastic. It's a winter voyage, a first, and we have a full agenda. We will sample plastic debris in a new season and in new places, and we'll scoop up a special species of fish for an important study. The study should help answer a pressing question: are gyre plastics potentially polluting the human food web? We know a section of the North Pacific Central Gyre contains, by weight, six times more plastic than zooplankton. And we know tiny broken-down plastic bits mimic both zooplankton and *their* staple food, phytoplankton. We've seen plastic bits glued to the insides and outsides of tubular salps and sticky jellies, both zooplankton species that ride surface currents, consuming whatever wafts their way. But what about the small fry fish that subsist on zooplankton, the so-called planktivores? They could well be carriers of plastic litter and possibly adsorbed toxics into the wider food web.

In scientific papers, you'll often see the terms *spatial* and *temporal*—qualifiers meaning, basically, the where and when of the study subject. This voyage will tackle new frontiers in both areas. Timewise, we hope to get a sense of whether plastic pollution is lighter or heavier this time of year, when the North Pacific is not very pacific. Spatially, we will journey farther north and west than ever before, toward the International Date Line, to an area north of the Northwestern Hawaiian Islands called the North Pacific Subtropical Convergence Zone (STCZ). This is where the meta-current ringing the entire North Pacific meets up in mid-ocean with the converging sub-gyres lying east and west of Hawaii, and with the smaller Alaska current on top. Within the STCZ lies a sort of marine median strip called the transition zone (TZ) where all these currents swirl past each other. Satellite images and low flyovers have given NOAA researchers confirmation of a "chlorophyll front" in the TZ, where phytoplankton looks to be especially dense. Like terrestrial plants, phytoplankton produce chlorophyll. And, like their counterparts on land, phytoplankton inhale CO_2 and exhale oxygen, to the great benefit of life on earth. In this same area they also saw concentrated debris, mostly fishing gear—net floats and buoys, plastic, of course—but also net clumps, including two monsters at least thirty feet wide. Dave Foley, a NOAA oceanographer, wants us to investigate this seeming correlation close up. We'll also trawl to see if microplastics, invisible in flyovers, are massing here along with plankton and "netbergs."

A captain never knows beforehand if a crew will mesh or clash, work or shirk. It hardly needs saying that forced coexistence on a twenty-five-by-fifty-foot slab of aluminum breeds instant familiarity and also tests character. This crew is gold. Jeff Ernst is a recent natural sciences graduate of the University of Hawai'i at Hilo. He proves a skilled deckhand with an uncanny knack for sensing and doing what needs to be done, and he wields his honed ceramic knife with excellent results in the galley. He gamely shinnies up the sixty-five-foot mast or hangs from the swinging boom as needed, earning the nickname "boat monkey." He's handy with a camera too. Joel Paschal could be Jeff's big surfer brother. We happily learn he worked in a bakery to help pay his way through college. He's also an underwater photographer and videographer who was hired and trained by NOAA to help with debris removal efforts in the Northwestern Hawaiian Islands. We call Herb Machleder "Doc." In fact, he's a retired UCLA surgeon as well as seasoned sailor and calming influence, especially for Anna Cummins, Algalita's new education coordinator and first-time deep-ocean mariner. Cummins is the type of person we need a few million more of in this world. Whip-smart, a Stanford grad, her environmentalism is fierce and core deep. Our fifth crew member is Marcus Eriksen, Algalita's science education and research adviser, a Ph.D. and bold eco-warrior. He's already navigated the Mississippi on a raft made of plastic bottles. On this voyage he and Joel hatch a more treacherous plan, a crossing from California to Hawaii in a "junk raft" made of plastic bottles, meant to draw attention to the ocean's plastic burden. Cummins and Eriksen met at my birthday party the previous May and soon became inseperable. She says if they're still talking by the end of the voyage, it might be the real thing.

The fish we hunt are myctophids, commonly known as lanternfish because they are bioluminescent. Myctophids glow, but they try to keep a low profile. Lurking in the oceans' "twilight" zone by day—the mesopelagic zone, 650–3,300 feet deep—they rise at night to feast on zooplankton. Lanternfish are among the planet's best-kept secrets. Many and small, rarely longer than several inches, they make up about 65 percent of deep-sea fish biomass. They've been known to blanket waters above the continental shelf so densely that oceanography instruments misread them as sea floor. Nevertheless, they are not considered a "schooling" fish, and thus escape commercial exploitation by fishermen with large nets, unlike anchovies and sardines. They populate every marine ecosystem as 254 distinct species, possibly more. At night, they literally rise and shine in a vertical migration which is the largest daily biomass shift on earth. The first time I ever saw a lanternfish was when we caught some in a night trawl during our first research voyage to the gyre. Lanternfish are food for creatures we eat—tuna, cod, salmon, and shark—as well as those we don't eat but do care about—whales, dolphins, pinnipeds, and penguins. The ingestion of plastic by fish is somewhat skimpily studied. Research in the 1970s and eighties found that whiting, pollack, haddock, and cod were at least occasional plastic eaters. Fish in the Irish Sea were thought to be hunting plastic trash tossed from the ferry between Wales and Ireland. A single pollack caught there contained five plastic cups. But the lanternfish will tell a different story about microplastics and the food chain.

We sail north to the transition zone (TZ) and do our trawls, setting aside samples for Dave Foley's study. Later we learn the samples were never analyzed, possibly due to a protocol glitch. This doesn't change the shocking reality of what we find. The assumption has been that plastic litter in the North Pacific clusters in two "garbage patches"—one located midway between Hawaii and California and the other between Hawaii and Japan. Both occur in "oligotrophic" areas where life-forms are comparatively scant. This might have meant polluting plastics were safely sequestered from more "productive" parts of the ocean that teem with marine life and fishing vessels, somewhat shielding the food chain from contamination. What we find in the transition zone shatters all assumptions. Our TZ trawls are the worst—the most plastic choked—we've ever seen. A distressed Cummins notes in the daily blog:

Here we are finding alarming quantities of plastic—more than we have ever found before—in an area of tremendous biological richness and commercial significance. We're finding the highest levels of pollution in a highly productive zone. The significance of this is far greater than people may have realized.

NOAA's study identifies this hot spot "as an important migratory and foraging habitat for a number of apex predators" and reports, alarmingly, that pelagic (open ocean) animals were "preferentially foraging" where marine litter was most concentrated. In other words, animals were targeting this place in the ocean where both plastic and food abounded.

With conditions favoring us, we leave the TZ and head south and east into the northeastern gyre, capturing lanternfish in seven trawls, six most productively at night, 670 fish in all. We store them in jars of formalin for later processing by Christiana Boerger, Algalita's surfing ichthyologist. Her findings will get the attention even of skeptics who doubt microplastic debris is harmful.

As I'd already learned while researching my first paper, there's no shortage of plastic ingestion studies. The first ones date from the early days of consumer plastics. They focus mainly on seabird populations ranging from the Arctic Circle to remote islands dotting the margins of Antarctic waters. Some species lend themselves more readily to study than others, but it's almost distressing how easily scientists find stranded marine creatures—not only birds but endangered sea turtles and an array of cetaceans—to examine or necropsy. The findings have been mostly grim.

Seabirds, mostly albatross, are what the NOAA flyovers spot in the transition zone. We see them too. Our northerly voyage to the TZ skirts the Northwestern Hawaiian Islands, where our 2002 trip took us. We glimpse the islands but cannot moor without prior permission. But we know they are home base for dominant transition zone forage feeders. If there is a poster child for plastics ingestion, it is, literally, the Laysan albatross. Who hasn't seen the stark images of dead, decayed Laysan albatross chicks stuffed with plastics?

Now the catastrophic March 2011 tsunami in Japan compels a reframing of this discussion in light of its savage effects on the main Laysan albatross habitat. It was good news in 2010 when the International Union for the Conservation of Nature "downgraded" the Laysan albatross from "vulnerable" to "near threatened." "Near threatened" is not exactly a free pass. It means the species "may be considered threatened with extinction in the near future" and warrants close monitoring. Between 1992 and 2002, the Laysan population suffered a 30 percent decline. Thereafter, its population leveled off, largely due to improved practices in the U.S. longline fishery. Then in March 2011 came news of tsunami devastation at the Laysan albatross breeding colony on Midway and elsewhere in the Northwestern Hawaiian Islands, with a preliminary figure of 110,000 chicks, nearly a quarter of all the hatchlings, drowned or swept out to sea. This alters the terms of any discussion about Laysan albatross and their unfortunate habit of feeding plastic to their chicks. Not only will there be fewer chicks to feed, but ocean debris will be vastly increased and all the more unavoidable for years and possibly decades to come.

Midway is not only where 70 percent of all Laysan albatross nest; significant populations of black-footed albatross, boobies, petrels, and shearwaters breed there as well. Before the tsunami, so many seabirds roosted on Midway—about two million in all, on an island less than twice the area of Central Park—that a flight service to the island canceled its contract due to bird-related hazards. In the midst of this seeming avian abundance, the shocking fact is that around 100,000 Laysan albatross chicks were dying each year under normal circumstances, an estimated 40 percent, or 40,000, from plastics ingestion.

Laysan albatross individuals have been known to live up to sixty years, including a female affectionately named Wisdom who nests on Midway and produced a viable egg in 2011, making the record books. She still bears her original tag. Albatross mate for life and breed once every two years, starting around age six. Once a chick fledges—first takes flight—at about six months, it heads out to sea, there to stay several years before returning to its place of birth. Mating occurs a few years later. On land, the albatross is awkward, waddling, roosting, and needing runway space for a galloping takeoff. Which is one reason it came close to total extinction in the nineteenth and early twentieth centuries, when millions were clubbed by feather hunters and poachers, mostly from Japan. The birds made easy prey, being ill equipped for quick getaways and lacking the instinct to make them. But once airborne, the Laysan albatross rules the skies. Its six-foot-plus wingspan and deep knowledge of air currents enables foraging missions across thousands of miles of ocean. What defines the albatross, and other seabirds, as "legitimate marine organisms" is their diet, strictly ocean sourced.

Midway albatross parents stake out a small patch of land for their nest—preferably on sand, sometimes atop plastic trash. Parenting is a shared effort. The egg laid, mom and dad jointly incubate it for the requisite seventy or so days. Then they brood, keeping the chick warm and safe for a couple of weeks, feeding it lightly. When the chick is able to thermoregulate, feeding in earnest begins. The parents take turns heading out to sea and bringing back meals. Sustenance consists of pre-digested catch in an oily slurry, which is regurgitated into the waiting chick's gaping beak. Albatross are piscivores (fish eaters) but also scavengers. They skim food items from the ocean surface, keeping a keen eye out for favorites: octopus, squid, krill, sardines, and leftovers from other predators (otherwise known as offal). A special delicacy is flying-fish eggs, typically deposited on a floating slab of debris, now often plastics. Older albatross like Wisdom would remember when lipstick-red flying-fish eggs would usually be draped across bobbing chunks of pumice or wood.

Barring tsunamis, all should be well. The days of mass slaughter are over. The albatross boasts all the legal protection the federal government can muster. Albatross parents are exemplary in their commitment and care. But the piscivores have become plastivores. As albatross luck would have it, plastic trash and legitimate food items share crucial characteristics. They tend to be shiny, colorful, and buoyant, and sized for snatching by the bird's nearly foot-long beak. Given new plastic inputs from the Japan tsunami, the albatross will be hard pressed to chance on natural food in the overfished oceans. Moreover, plastic doesn't put up a fight. Having evolved as an unfinicky "surface feeder," the albatross, a visual hunter, chronically falls for a literal game of bait and switch. A mere two generations ago, whatever was out there would have been a safe bet to eat or take home to Junior. Now NOAA's observations of the Convergence Zone suggest seabirds may be heading to this area *because* of easy plastic pickings.

It's a given that a chick will be fed plastics. If it's lucky, its parents will feed it enough natural food and not so much plastic that it winds up resin-stuffed, dehydrated, and starving. If it's lucky, none of the plastic pieces will be sharp enough to perforate an organ or block the digestive tract. If it survives five months, it will attain a key albatross rite of passage: regurgitation of its first bolus. The bolus is a sort of avian fur ball composed of indigestible squid beaks, pumice, fish scales, wood bits, feathers, and now plastics. The ability to regurgitate protects the adult albatross from most of the harm plastics can cause. But the chick is in a race against time.

Cynthia Vanderlip has seen firsthand the slow, sad ebbing away of life in Laysan albatross chicks—the size of geese, by the way—on both Kure Atoll and Midway, and she's performed necropsies. Invariably, she finds a jumble of synthetic resins, mostly fragments, but also things like bottle caps, disposable lighters, toothbrush handles, and toy figures. Midway staff say adult albatross may be hauling in five tons of plastic each year, culled from the ocean surface.

Beth Flint is a seabird specialist who is supervising wildlife biologist for the U.S. Fish and Wildlife Service on Midway. In her public presentations, Flint says, "Starting in August and continuing through winter, chick carcasses are a common sight. And as they decay, you can see that virtually all of them have plastic debris in their guts." But this good scientist doesn't see in these remains conclusive proof of death by plastics, given other threats, such as lead paint on old buildings and parental abandonment, which would be due in part to longline "bycatch" mortality. Asked if there's a pre-plastics baseline rate for albatross chick mortality, she says no. Indeed, Flint notes that by providing more floating platforms for flying-fish eggs, the albatross food supply could be improving. And yet it's hard to imagine that fish eggs served to chicks on plastic rafts are helping the species. We capture such a meal in a morning trawl on the 2008 winter voyage. It's a mass of transparent eggs clustered like little bubbles in a knot of fishing line. These eggs also happen to be a favorite food of the black-footed albatross, which is in greater decline than its Laysan cousin.

In another fine example of science unhinged from public awareness, the study of plastics ingestion by the albatross (not to mention dozens of other seabirds—44 percent of all species) goes back nearly fifty years. The first was in 1963. This study found that 73 percent of Laysan albatross at Pearl and Hermes Atoll, another of the Northwestern Hawaiian Islands, had "swallowed" plastics. But in these early days of consumer plastics, the greatest number of particles they found in any bird was eight. The next major albatross study came in 1983. By this time plastics were observed in 90 percent of dead Laysan albatross chicks, and the average weight of ingested plastics had risen from 1.87 grams in 1963 to 76.7 grams—nearly three ounces—a jump of 3,000 percent. A 1997 study was coauthored by Theo Colborn, the wildlife biologist famous for linking synthetic compounds with endocrine disruption in the book *Our Stolen Future*. At that time, plastics were found in 97.6 percent of chicks sampled. The authors concluded that the concentration of plastic debris across the surface of the north-central Pacific Ocean was trending upward. This was the year I sailed by chance through the Garbage Patch. Seabirds' stomach contents had already become barometers of ocean pollution, but who knew? By this time, MARPOL Annex V had been in force for nearly a decade.

The disposable lighter is an albatross favorite. The glint of metal and kaleidoscopic colors would appeal. Vanderlip observes, "Color is of interest to albatross. There aren't any studies, but I notice that they peck and bite at my colorful clothes and shoes. They eat colorful crustaceans, so they favor reds and blues." Over a two-month period on Midway, volunteers collected 1,310 lighters from the nesting grounds, many thought to be tossed by fishermen. Other artifacts recovered from dead chicks have included vintage plastic from a World War II fighter plane (the oldest identifiable pelagic plastic), toothbrushes, combs, beads, plastic buttons, checkers, golf tees, dishwashing gloves, and Magic Markers. And the most common debris object of all: plastic bottle caps, made of durable polypropylene, rarely recycled, likely to outlast us all.

No one is keeping statistics on bottle caps per se, but their numbers can be extrapolated from bottle production figures. For example, the Container Recycling Institute reports that each American consumes on average 686 single-serve beverages a year. That's 215 billion containers. Of these, 75 billion or so were PET or PE bottles. About a quarter at most of these bottles would be recycled, most without the caps. Now consider how many additional caps are made for products like pharmaceuticals and

supplements, shampoo and conditioner, sunscreen and skin lotions, liquid soaps and cleaning products, ketchup and pancake
syrup. Even a tiny percentage of all those caps rolling into the
oceans each year, and accumulating, will begin to rival natural
food in the pelagic layer.

As to science's lack of "quantitative proof" of the number of
Laysan albatross chicks killed by plastics, Cynthia Vanderlip
says, "At this point I think we have to ask ourselves if we need
scientists to tell us everything. I don't think I do . . . I think we
need to err on the safe side."

I often think of *Alguita* as voyaging like an albatross. When the
wind is fresh, albatross soar on fixed wings. When it is calm they
have to flap. When the wind blows, we sail. When it is calm our
diesels "flap" our propellers and keep us going. We often sight
albatross on our voyages, but the thrill never fades, even if they're
clearly more interested in our fishing lures than us. An endangered black-footed albatross once hoved into view above our wake,
and the crew raced to the aft deck to have a look. I said, "I hope
she doesn't go for the lure." And then she did, and got hooked.
We quickly reeled her in. It was Joel Paschal who unhooked and
cradled her a bit to soothe her—it made a nice photo—before
tossing her back to the sky. These birds might seem reckless, but
they only do what comes naturally.

In 2009, Algalita researcher Holly Gray studied the stomach
contents of forty-seven bycatch adult Laysan and black-footed
albatross, recovered and frozen for research purposes by NOAA
at-sea observers on fishing vessels. Heretofore, only chick guts
and regurgitated boluses of adult albatross had been examined.
While noting several potential confounding factors—the birds'
trauma may have prompted regurgitation prior to death, and
natural foraging habits might have been altered by the presence

of fishing vessels—Gray found that 83 percent of the Laysan and 52 percent of the black-footed albatross contained plastics. Most plastics in the Laysans were fragments; in the black-footed, fishing line. But neither came close to matching the volume of plastics found in the typical chick bolus or gut. The adult albatross easily regurgitates plastics, but young chicks can't. The findings were significant for showing how pervasive plastic ingestion is among these surface feeders. They also support albatross tracking studies showing the Laysan heading north to the plastics-dense Convergence Zone and black-footed heading for the U.S. West Coast, where plastics are more dispersed in rougher waters.

A 2002 study in the Netherlands found that 80 percent of large floating plastic pieces washing ashore had been pecked by seabirds. Culled from various studies are these figures for percentages of several bird species found to have ingested plastic: puffins, 95 percent; blue petrel, 93 percent; northern fulmar, 80 percent. A recent study of blue petrel chicks at South Africa's remote Marion Island showed that 90 percent of chicks examined had plastic in their stomachs, apparently fed to them by their parents. This puts blue petrels in dubious company with the albatross, to whom they're related. Unlike the Laysan albatross, the somewhat smaller petrel dives up to twenty feet deep to grab prey, which raises questions about submerged plastics as possible prey items.

Seabirds happen to be more easily studied than other marine species because they divide time between land and ocean and tend to be surface feeders. But many other species are equally prone to mistaking plastic debris for food and suffering the consequences. Any doubt as to the ubiquity of marine plastic should be demolished by the simple fact that plastics are found in the smallest filter feeders on up to the ocean's, and the earth's, largest creatures, whales. Among the most vulnerable are endangered sea turtles. A 1997 study by the Marine Mammal Commission looked at data culled from necropsied stranded animals, any type of marine creature that had wound up lifeless on a shore. Only two types had never presented evidence of plastic ingestion: sea otters and crustaceans.

Sea turtles are a special case. Marine turtles evolved millions of years before man, with systems successfully geared to survive in a world without Homo sapiens. Now that we're here, five out of seven sea turtle species in U.S. waters are endangered, the others threatened. Research tells us primary threats to marine turtle survival are fisheries bycatch, vessel strikes, predation by man and other creatures, both marine and terrestrial, and a relatively new issue, a herpes-related tumor-causing disease called fibropapillomatosis. Given this suite of challenges, adding in lethal peril posed by marine litter hardly seems fair. In 1985, George Balazs, a NOAA turtle specialist posted in Hawaii, confirmed seventy-nine cases of turtles with guts filled with various types of plastic debris. Upon conducting necropsies of dead turtles, researchers in the Mediterranean found that nearly 80 percent had ingested marine debris, mostly plastic. Slightly lower percentages were found in studies of stranded turtles in Brazil and Florida. Sea turtles are famously prone to mistaking plastic shopping bags for one of their favorite foods, jellyfish. A 1988 study cited a legendary sea turtle recovered in New York and necropsied: 540 meters—nearly a quarter mile—of fishing line was extracted from its stomach and esophagus.

Mortality statistics for marine turtles are mostly based on adult turtles that perish, then wash ashore. But sea turtles have a curious developmental quirk that renders an entire and highly vulnerable group—hatchlings and yearlings—almost impossible to study. French Frigate Shoals happens to be the central nesting place for the threatened Hawaiian green sea turtle. Upon hatching, these little guys dig their way out of their sand nests and head for the water, guided by the moon's reflection. Led by instinct, they paddle toward the open ocean, there to spend two "lost years"—this being their pelagic stage—before setting scaly foot on land again. They elude study but face a multitude of threats. Little is known about their travels during these years, but inevitably many would encounter plastic morsels in the neuston layer they navigate.

When we moored at French Frigate Shoals in 2002, it was green sea turtle hatching season. We volunteered to help U.S. Fish and Wildlife Service personnel steer hatchlings out of the lagoon and into the open ocean. Typically brisk currents cause many to be washed up on the atoll's reefs. Others are eaten by the ulua or giant trevally, big members of the jack family that patrol the lagoon. So we took the babies on board the *Alguita* and motored them to safer waters outside of the reef. They were so cute with their little flippers flailing, straining against the current. To think many of these hatchings wind up in the gyre ingesting plastics is more than a little distressing. In my PowerPoint presentations, I show a photo of a baby turtle recovered by an Australian crew. It's not readily apparent that the little creature is dead, but it is, killed by two plastic flakes lodged in its pylorus, blocking elimination. Plastic pieces that would be passed by a larger animal can kill young ones.

Like plastic bags, balloons and balloon bits seem to exert a special appeal to the sea turtle. We've certainly come across our share of balloons in near-shore investigations, and even found the foil variety hundreds of miles offshore. On a voyage one brisk blue morning in February 2010, following several days of rain in Southern California, we cast off from Long Beach and headed out to see what manner of new debris has been swept from land into the shore waters. The sights that day included a windrow made almost entirely of plastic straws, a small pod of dolphins, and California sea lions basking on a buoy—one of them sporting a noose of plastic fishing line. Holly Gray, our bird specialist, blogged the day. The "strangest item," she wrote, was

> a balloon. Balloons unfortunately are an extremely common sight on the water. . . . People throw parties and release their balloons into the air. We saw balloons of all shapes and colors today but this one was different. We could see this bright pink balloon from quite a distance. When we got closer Captain Moore skillfully captured it with the boat hook. Pink, shiny and adorned with a picture of Hannah Montana the balloon read "Let's Rock." And sure enough, hitting the balloon with the boat hook caused it to launch into song from a small speaker embedded inside.

This was a so-called foil balloon made of nearly indestructible b/o-PET (biaxially oriented PET) metalized film. DuPont discovered and patented this material in 1954 under the name Mylar, but rarely is real Mylar used to make these cheap but durable novelties. Indeed, virtually all are manufactured in China. Inflated with helium, they may delight the birthday girl, but escaped balloons are a proven menace to power grids and the marine environment. In 2008, a bill to ban the foil balloon passed one house of the California legislature but died a few months later after the balloon industry—yes, there is a balloon industry—launched a full-court press in opposition. The stats they cited were eye-popping. A report in *The Wall Street Journal* shed light on just how vast this industry is, if their self-reported statistics are accurate. A spokeswoman for the Balloon Council told the *WSJ* that "45 million foil balloons [are] sold in the state a year, selling on average for just over $2 a pop. When combined with floral arrangements or teddybear gifts, the gross hits $900 million. The state would be losing $80 million a year in sales tax if the ban goes through."

So it turns out the balloon lobby is highly organized and engaged in something like perpetual warfare with environmental advocates who would like mass balloon releases to be banned. According to Clean Ocean Action, a group serving the mid-Atlantic seaboard, "There is a greater than 70% chance that airborne balloons or their fragments will end up in the oceans and harm marine life. Scientists who work with stranded whales, dolphins, seals and sea turtles have found balloons, parts of balloons and balloon string in the stomachs of many of these dead animals." The group goes on to report that, in 2003, volunteers collected 4,228 Mylar and latex balloons from New Jersey beaches alone.

These figures stand in stark contrast to "the facts" published on the Balloon Council's website, Balloonhq.com. The council cites beach cleanup data from the 1990s that found that balloons "accounted for .64 percent of the total debris collected." The site claims that data from subsequent beach debris studies is pending. Only in their dreams is such data still pending. That data, and many more recent data sets, are in and readily available. The Marine Conservancy Society, based in the UK, where balloon releases thrive, tabulates beach debris and denounces helium balloons and balloon releases. As of 2008, its figures showed a 260 percent increase in balloon-related beach litter over a decade. But here's what the Balloon Council website claims: "Bottom line— balloon litter has never been a significant part of the list of debris and it continues to drop."

Even the balloon lobby condemns mass releases of foil balloons. But it staunchly defends "professional" mass launchings of latex balloons. Latex is made from natural rubber that is biodegradable. When latex helium balloons climb five miles in the sky, they begin to oxidize, embrittle and expand, at which point, the Balloon Council reports, they shatter into "spaghetti-like" pieces that fall to earth and decompose at about the "rate of an oak leaf." Taking up the gauntlet, a group in England studied oak leaf degradation. In certain conditions, it found, an oak leaf can take up to four years to biodegrade. The Balloon Council says about 10 percent of balloons descend intact. Environmental groups tend to put the figure higher.

Our friend the plastics researcher Anthony Andrady has weighed in on the issue. He made this determination:

Latex rubber balloons are an important category of product in the marine environment. Promotional releases of balloons that descend into the sea pose a serious ingestion and/or entanglement hazard to marine animals. Based on the fairly rapid disintegration of balloons on exposure to sunlight in air, the expectation is that balloons do not pose a particularly significant problem. In an experiment we carried out in North Carolina we observed that balloons exposed floating in seawater deteriorated much slower than those exposed in air, and even after 12 months of exposure still retained their elasticity.

Balloon releases tend to mark occasions like graduations, weddings, and funerals, and to support noble causes and store openings. The soaring balloons are meant to embody hopes, dreams, and aspirations—headed for blessings on high. But what goes up must come down. In the early 1990s, Peter Lutz, noted sea turtle biologist based in Florida, conducted an experiment with sea turtles. He found that when turtles were offered a choice between pieces of clear plastic and brightly colored latex, they almost always motored toward the latex bits. He also found that a hungry turtle deprived of natural food will ingest balloon pieces regardless of color. And, of course, the ribbons tying all balloons are non-biodegradable plastic.

The deeper one delves into plastics ingestion, the more one feels stuck on a horrific carnival ride that turns corner after corner, each revealing a fresh diorama of gruesome death. It's no accident that most ingestion literature involves seabirds and turtles, which evolved as opportunistic surface feeders. But by virtue of its sheer volume, plastic debris has become harder to avoid for more discriminating marine mammals. In August of 2000, a Bryde's whale—among the smaller of baleen species—stranded on an Australian beach, alive but clearly suffering. Local rescue folk erected a tarp shade over the creature and doused it with seawater. A video painfully records its final moments, thrashing followed by surrender. The necropsy astounded. From the animal's gut spilled nearly six square yards of compressed plastics, mostly shopping bags.

A thirty-seven-foot-long "sub-adult" gray whale was stranded on a Puget Sound beach in March 2010. Whale strandings occur seasonally, if not frequently, as grays head south on their annual migration. This necropsy catalogued the following items taken from the whale's abdomen: sweatpants, a golf ball, surgical gloves, small towels, plastic fragments, and twenty plastic bags. The man-made contents made up a small percentage of the mostly organic fifty gallons of material found in the whale's stomach. But any amount was considered significant, and "unusual," because the gray whale is a benthic, not surface, feeder. This creature scoops up gallons of sludgy material from the seabed, then expels it through the natural sieve of its keratin baleen. It's what would happen if you had a mouthful of vegetable soup and squirted out the broth through your teeth (not recommended). What remains for swallowing should be krill, octopi, crustaceans, and other favorite bottom-dwelling foods, not plastic bags and gloves. While this death could not be directly attributed to plastic debris, those of two pygmy sperm whales that were stranded in the northern Gulf of Mexico in the 1990s were. One had beached itself alive on Galveston Island but died eleven days later in its holding tank. The necropsy showed its first two stomach compartments were lethally and "completely occluded by various plastic bags." The other whale shared a similar fate. The United Nations' Convention on Migratory Species catalogs threats to border-crossing species. Its latest inventory, published in 2010, cites among threats to migratory cetaceans multiple confirmed cases of consumer and fishing-related plastics ingestion.

And while consumer plastics are the focus of our efforts, fishing debris still rankles. Dr. Frances Gulland is the director of veterinary services at the renowned Marine Mammal Center on San Francisco Bay. This impressive facility treats about a thousand seals and sea lions each year and conducts research as well. Gulland's expertise extends across marine mammal species, and, as of 2011, she is a presidentially appointed marine mammal commissioner, one of only three. The majority of the center's cases involve disease, malnutrition, and "anthropogenic" harm,

typically entanglement and injury inflicted by humans. But
Gulland encountered something well out of the norm when she
joined a pathology team dispatched in 2008 to the sites of two
sperm whale strandings—barely a month apart in late winter, on
Northern California beaches. She kindly forwarded the resulting
report, which I summarize here.

The whales were necropsied where they were stranded. The
first appeared to be in good shape, with no signs of emaciation
or scarring. When the abdominal cavity was opened, the team
saw "a large mass of compacted netting" protruding through a
bloody perforation in the whale's abdominal wall. The cause of
death was "presumed" to be "gastric rupture" caused by impacted debris. The second whale was a male, forty feet long, in
"poor nutritional condition" and marked with welts and scarring
consistent with entanglement. This whale's stomach was intact,
but the third chamber was impacted by a large amount of netting, line, and bag scraps. The animal had seemingly starved to
death. The scientists had debris contents analyzed at the Hum-
boldt State University Vertebrate Museum. The largest piece of
netting recovered was about forty-five square feet. The second
whale would have swallowed more than two hundred pounds of
debris, the first far less. Most of the pieces of netting had a
"twisted" cord hallmark that indicated Asian origin. Some appeared well worn. The report speculated that natural prey entangled in the nets might have seduced the whales, and that the
ingestion likely occurred during winter in the North Pacific
Central Gyre.

A dolphin ingestion story printed in *Audubon* magazine has
a rare happy ending but also serves to show how intelligent but
understimulated animals can be lured by plastics. Two star bot-
tleneck dolphins at the Royal Jidi Ocean World in Fushun,

northeast China, discreetly amused themselves by "nibbling" plastic from the sides of their pool. When they lost their appetites and "became depressed," veterinarians discovered the problem. Surgical tools proved unequal to the task of extracting the plastics orally. So the world's tallest man was summoned, a Mongolian herdsman named Bao Xishun. At seven feet, nine inches, Bao has extraordinarily long arms. Staff used towel slings to hold open the dolphins' mouths and prevent bites. Bao was able to reach down the mammals' long throats and pull out handfuls of plastic shards. The dolphins reportedly fully recovered.

The plague of plastics ingestion isn't confined to the marine environment, and it's made for unlikely alliances. A crusty, politically conservative rancher on the Hawaii's Big Island has grazing pastures within a few gusts of a garbage "transfer station." He joined forces with environmental groups to attempt a plastic bag ban on the island. What got the ball rolling was his letter to a councilperson stating that several of his calves had choked to death on errant bags. He described calves as especially vulnerable, given their natural curiosity and playfulness. Unfortunately, the resulting bill was twice shot down by a conservative council—supported by pro-bag retail lobby efforts—but it's expected to prevail in time. News of local bag bans—including one in Los Angeles County and my own city of Long Beach, for which I gladly testified—are on the verge of becoming commonplace. Now Italy has banned the lightweight shopping bag.

In the United Arab Emirates, a German-born veterinarian, Ulrich Wernery, is science director of the Central Veterinary Research Laboratory in Dubai. According to Gulfnews.com, Wernery made a gruesome discovery in 2007 when he investigated an outlying valley reputed to be a dumping ground for dead camels and livestock. There he found thirty carcasses. Necropsies revealed calcified balls of plastic bags and ropes in camels' stomachs, one weighing well over one hundred pounds. He now believes one in three camels in the UAE dies from plastics ingestion. On average, individual Dubai residents generate more than a ton each of waste every year, one of the highest per capita rates anywhere in the world. But plastics consumption has far outpaced development of waste disposal systems. Wernery told *Gulf News*: "This is the worst environmental threat facing this country. Death of our animals from plastic is reaching epidemic proportions . . . but people won't do anything about it." Besides camels, animal ingestion victims in the UAE include sheep, goats, gazelles, ostriches . . . even houbara bustards. Published photographs show donkeys and goats picking through heaps of plastics bags, looking for food scraps.

Responding to the crisis, Wernery founded the Emirates Environmental Group. His efforts are bringing about reform, including construction of a world-class recycling facility near Dubai. The federal government has decreed that all plastic bags made anywhere in the UAE must be biodegradable by 2013. But in rural areas, vegetation for grazing is sparse and plastic trash still can be as thick as in Times Square on New Year's.

The sacred cows of India are likewise falling prey to plastic trash. India's new prosperity includes an enormous surge in disposable plastic containers and packaging. India is the third-largest consumer of plastics and its chemical industry is largely focused on polymer production, for export and internal processing, by forty thousand plastic product converters. According to Indiatimes.com, plastic waste generation in India is estimated at 4.5 million tons per year and households use ten to twelve plastic bags each day. Several Indian states have banned or restricted use of the plastic bag. As in other emerging economies, waste man-

agement infrastructure lags behind burgeoning plastic waste generation. Cow deaths from plastic bag ingestion are estimated at
one hundred per day in the state of Uttar Pradesh. Cows in India
can be something like opportunist feeders in the marine environment, owing to some dairies freeing their stock between milkings
to forage in the streets. It's a costly economy. The cows graze
trash heaps, sniffing out food residue on bags that then become
meals. A cow rescue worker in the capital city of Lucknow described the classic bovine victim: an emaciated cow with a bloated
stomach. In one widely reported case, a dead cow was found to
harbor a clot of plastic weighing seventy-seven pounds. Not as
bad as some of the Dubai camels, but still horrendous, and hor-
rendously needless.

Pinnipeds—seals and sea lions, among a few other species—
are carnivores and smart enough to tell the difference between
favored edibles—schooling fish, crustaceans, and cephalopods—
and plastic. Debris entanglement is their thing, not mistaking
plastic for real food. But I learn of a case that surprised even
seasoned marine mammal rescuers when I schedule a visit to the
Marine Mammal Care Center at Fort MacArthur, in nearby San
Pedro, in late 2010. I'm greeted by Dr. Lauren Palmer, the veterinarian in charge. We stroll among the tanks, where patients
appear to be happily and healthily frolicking in the water and
*ork*ing their hearts out. There are few patients—only nine—at
this time. She says the majority of injured animals she sees during
lulls suffer from commercial fishing interactions, what I also
heard next door at International Bird Rescue. These include
stabbing and gunshot wounds that tend to be inflicted by fishermen who view pinnipeds as competition. They are well aware
harming protected marine mammals is a federal offense unless
rigorously justified. She shows me some photos. One is of a sea

lion that became trussed in monofilament fishing line as a pup. As it grew, the line's grip tightened, incising its skull and creating a crevasse-like malformation across the head. Its neck was similarly creased by the line's noose action. This animal survived but, like many, wasn't fit enough to be returned to the wild. Palmer tells me the center places nearly all its handicapped patients in marine parks, where rehabilitated animals become visitor favorites and actually boost attendance. She also says she expects more patients—maybe hundreds more—during the spring pupping season. This is also when the center sees a sharp spike in domoic acid toxicity. Domoic acid poisons coastal water when a tiny aquatic plant, a diatom named *Pseudo-nitzschia,* suddenly proliferates or, as they say, blooms. It makes coastal creatures very sick and is the bane of marine animal rescue centers. How do these blooms relate to plastics? Bear with me.

Most people know them generically as red tides. But marine protection agencies call them harmful algal blooms, or HABs. It's a bigger subject than most realize, reaching far beyond periodic warnings about tainted shellfish. HABs cost billions in economic loss to fisheries and coastal communities and they cause many of the mass strandings of dolphins, pelicans, and other creatures that we hear about on the news. The EPA has formed a task force to investigate the phenomenon and is aggressively funding research. The bad news is this: HABs, a catchall term for a varied group of toxic and nontoxic algae types, are growing in size, intensity and frequency, all over the world. I check in with my coastal waters guru, Steve Weisberg at SCCWRP, and he confirms how seriously the issue is taken and how much is still to be learned. The basics are known: the phenomenon is seasonal, occurring from late winter through late spring, when coastal waters are cool, and it correlates with seasonal "upwelling," an oceanographic dynamic that brings enriched sediments to surface waters. There appears to be a connection between HABs and runoff from land, mainly sewage effluents and fertilizer, which contain minerals and chemicals that enhance both marine and terrestrial plant growth. Groups like SCCWRP and the Caron Laboratory at the University of Southern California are working together to determine if HABs can be predicted and if they can be controlled or mitigated. Marine animal rescue groups brace themselves for HABs season, not knowing if it will be a bad one or mild. Spring and summer of 2007 brought the worst HABs event ever seen in Los Angeles and Long Beach harbors, and one of the most virulent ever measured.

Palmer remembers it well. More than a thousand sea lions and seals were brought to the center over several months, and the bird rescue center was likewise inundated, with animals poisoned by HAB-tainted fish. The sickest require euthanizing. Domoic acid is an insidious biotoxin, a neurotoxin that causes seizures and behavioral abnormalities as well as organ damage and sometimes paralytic death. Pregnant pinnipeds are disproportionately affected because they're eating for two, in this case consuming infected shellfish and planktivorous fish like herring and sardines.

Entire ecosystems are disrupted as the toxin works its way through a local food web. Poisoned pelicans have been known to seize as they fly, landing with a thud on windshields. Sea lions have been found wandering on busy highways, miles from home. Those whose seizures can't be controlled in rehab must be euthanized; their seizures would cause drowning in the natural habitat. Humans who consume tainted seafood have been known to die. And the connection to plastic debris?

Though I know plastic ingestion is rare among pinnipeds, I ask Palmer if she's ever seen such a case. In response, she brings up a photo on a computer. It's of a sea lion with, unmistakably, a plastic bag protruding from its lower abdomen, as if being delivered by Caesarean section. She tells me this was a female rescued at Malibu and brought to the center in 2007, the year of the very bad HAB. When the surgery was over, thirteen plastic shopping bags had been pulled from her stomach. She survived another forty-one days. Dr. Palmer assures me that she had multiple health problems, including domoic acid toxicity, several of which could have sealed her fate. But Palmer thinks the neurotoxic effects of domoic acid might have spurred the sea lion to eat bags when normally she never would have. The patient was emaciated, a symptom of plastics ingestion, and the bags in her gut would have inevitably killed her, even if she'd recovered from the poisoning. If the bags hadn't been where they didn't belong, adrift in shore waters, the Malibu sea lion would at least have been spared a little misery. Palmer said she'd like to see a study that investigates a possible correlation between domoic-acid-induced dementia and plastics ingestion.

On a hunch, I decide to have a look at HAB's research. I find a study from 2003, carried out in Mediterranean coastal waters of Spain. During one HAB outbreak, the lead scientist, Mercedes Maso of the Institut de Ciències del Mar (Institute of Marine Sciences) in Barcelona, recovered pieces of plastic litter from coastal waters and examined them under a microscope. She found HAB spores—in a "sticky" stage—attached to the plastic and theorized that plastic debris, propelled by currents, could spread toxic algae spores. She urged further study. My thoughts take a speculative turn. We know a number of synthetic chemicals associated with plastics—intrinsic chemicals as well as persistent organic pollutants they adsorb—have estrogenic qualities. Ethylene gas, the basic building block of the most prolifically produced single-use plastic, is a major plant hormone, applied commercially to hasten ripening in several fruit crops. Foamed polystyrene was tested as a growth medium for ornamental plants and found to be a significant growth booster. How about heavier plastics that have adsorbed pollutants and sunk to the seabed? Would they fortify what upwelling brings up from the benthos? Testing for bioactive effects on plankton by plastics seems worthwhile, I think. I contact a few HAB experts who seem intrigued. A young professor at the University of Hawaiʻi at Hilo plans to use his department's new electron microscope to check plastic debris for several toxic spores known to inhabit Hawaii's subtropical waters. In due time, there may be new data to share, though without experimental results, my thoughts can only be considered speculation.

THE TANKS-AND-SHORTS WEATHER of the middle gyre gives way to storm gear as we near the California coast in late February. The crew fantasizes about hot showers and fresh produce. We arrive back in port on February 23 after more than a month at sea. Anna Cummins now sports an engagement ring that Marcus Eriksen, a mechanical whiz who can make just about anything from anything, wove from rope fibers. She fancies she's the only woman ever to be proposed to in the North Pacific Subtropical Gyre, and on Valentine's Day to boot. We have jars of preserved fish for Christiana Boerger's dissecting microscope, and we're eager for her to get started. Will the 670 lanternfish we captured for her be plastics-free or plasticized?

It will take another year for each little fish to be opened and examined, the data to be compiled, and the paper written. We're thrilled when the *Marine Pollution Bulletin* deems it worthy after peer review. The significance of this study—the reason it stands out among the many scores of studies investigating plastics ingestion in marine animals—is that it addresses a new category of creature. Myctophids are "lower trophic" organisms, meaning they occupy a lower rung of the food chain, one that essentially exists to provide food for higher trophic animals. We have a pretty good idea of which creatures eat lanternfish, but this type of study can be tricky, given the vast arena the subjects occupy. King penguins eat them, we know. A 2006 study by French scientists confounded assumptions about dolphins' feeding habits and is germane to Boerger's work. Two dolphin populations, thought to be opportunistic feeders, were discovered to be selectively hunting and eating two varieties of myctophid. The reason is that lanternfish provide a greater "energetic load"—more calories—than other available prey, making for better bang for the hunting buck. A number of pinniped populations in Alaska, Mexico, and sub-Arctic waters are known to dine on myctophids. On Macquarie Island, near the Antarctic Circle, plastic was found in seal scat. This was seen as evidence that plastics moved up the food chain, quite possibly when seals ingested plastic-eating myctophids. A particularly fascinating, and relevant, study looked into hunting strategies of spinner dolphins off the coast of Oahu in 2008. The researcher tracked a group of dolphins as they herded prey into a tight circle and took turns feasting on their corralled quarry. What I found most interesting was that the hunt occurred at night and the prey was lanternfish, which, of course, vertically migrate to the ocean surface at night. Coincidentally, the number of lanternfish a dolphin requires to meet its daily calorie requirement is 650. This is almost precisely the number of lanternfish Boerger studied: 670. In light of this, consider her results:

What she found was that 35 percent of the sampled lanternfish contained plastic bits. The larger specimens contained more plastic. Cumulatively, the fish yielded a total of 1,375 plastic flecks, averaging one millimeter in size. A single fish, the record holder, contained 83 plastic fragments. The larger fish contained on average 7 pieces. The vast majority of the bits were fragments, with film and bits of line and rope together adding up to less than 6 percent.

Now we know, as if there'd been any doubt, that lots of different creatures accidentally eat plastics. We also know incalculable quantities of plastic litter—two hundred thousand homes and their contents were swept out to sea, for starters—were introduced to the Pacific Ocean by the March 2011 tsunami in Japan. Ocean currents and winds will widely disperse this debris throughout the pelagic habitat, and much will be mistaken for food and consumed. We have a sense of what kind of mechanical harm plastic litter can cause in the guts of living things. But we've been skirting around an important issue. It's time to consider whether the world and its inhabitants, including people, are being poisoned by plastics.

CHAPTER 13

BAD CHEMISTRY

Though the world does not change with a change of paradigm,
the scientist afterward works in a different world . . .
—THOMAS S. KUHN, *The Structure of Scientific Revolutions*

IN 2005, ALGALITA HELD a conference called "Plastic Debris Rivers to Sea." Naturally, the program focused on contamination of oceans by river-borne plastic debris. If I do say so myself, it was a marine debris extravaganza, featuring top names in the field and the latest research. The conference also marked the inauguration of a project that continues to this day, namely International Pellet Watch (IPW). IPW was something of a dream come true for one of Algalita's good friends, Hideshige Takada, a professor at Tokyo University of Agriculture and Technology. Takada coauthored the seminal 2001 paper that proved preproduction plastic pellets, nurdles, aggressively attract and adsorb persistent organic pollutants (POPs) from polluted seawater—and put a nice edge on my first research paper. That study planted the idea that plastic pellets could work as monitors of environmental pollution. When I invited Takada to present at the conference, he asked if he could roll out his concept for IPW at this ideal venue. But of course! The idea would be to solicit beach-collected resin pellets from around the world and to test them for a number of POPs, including some newer ones. It seemed an ingenious and economical way, in Takada's words, "to understand the global distribution of the pollutants in marine plastics." Typically, monitoring efforts involve water or wildlife sampling, both of which entail considerable technical coordination and expense. His assumption was that pellets would register local seawater pollutants as they passed through contaminated waters on their way to shore.

But as the Japanese research team had found five years earlier, their plastic pellets give almost as good as they get. Their measurable toxics were both sponged up from surrounding waters and released from the plastic material itself. Manufacturing plastics, and getting them to do all the remarkable things expected of them, entails modification by a welter of chemicals that catalyze, lubricate, stabilize, harden, soften, strengthen, rubberize, colorize, texturize, flameproof, germproof, heatproof, and prevent oxidation. According to some estimates, each element of a product such as a plastic baby bottle, with its rubbery nipple, hard plastic collar, and clear bottle base, could individually contain dozens of chemicals, none of which the manufacturer is required to disclose thanks to trade secret protection. The IPW project will test for some of these intrinsic chemicals as well.

By 2010, wielding stainless-steel tweezers or well-washed hands, volunteers in twenty-three countries on every continent gathered at least one hundred, preferably two hundred, pellets from fifty-one coastal sites. The older and yellower the pellets looked, the better, because those had been exposed longer and would yield more telling results. Volunteers were instructed to use stainless-steel tweezers or well-washed hands for gathering, and *not* to wash the pellets and *not* to wrap them in plastic for mailing, but instead to use foil or paper, note the GPS coordinates of the recovery site, and send them by airmail to Tokyo. The findings thus far: San Francisco pellets are the most polluted, followed by Tokyo and Boston. The more populous and industrial the surrounding community, the higher the rates of alphabet-soup toxics: PCBs, DDT, PBDEs (polybrominated diphenyl ethers, used as flame retardants), PAHs (polycyclic aromatic hydrocarbons—compounds released by incomplete burning of carbon-containing materials like oil, wood, tobacco, garbage, and coal—of which there are many thousands), and plastic additives, including the notorious BPA. The cleanest: pellets from Thailand, Costa Rica, Hawaii, and other less industrialized areas. We offered plastic fragments from our gyre cache, having no clue how they'd compare with coastal samples. The surprising results did nothing to allay our worries about the marine food chain, but we will save them for later, after we've delved deeper into the discomfiting properties of these synthetic potions.

Two big points. One: Takada's tests compared the toxicity of the two dominant plastic types: polyethylene and polypropylene. The most contaminated by far was polyethylene. Unfortunately, polyethylene is also by far the most abundant type of plastic used to make shopping bags and most plastic packaging, and it's about 75 percent of the plastic in our ocean samples. Two: not only does pellet contamination reveal what's in nearby seawater, it also appears to correlate with sediment contamination. Sediment is a mix of inorganic silt, sand, and metals and organic remains, droppings, and seepings. As such, it's oily and therefore binds with oil-soluble contaminants. Thus, in polluted coastal areas, marine creatures have multiple opportunities to encounter man's toxicants, from seabed to ocean surface. Upwelling from the toxic seabed provides additional opportunities for plastic debris to adsorb and concentrate these toxicants, a recipe for poison plastic pills that are up to a million times more toxic than surrounding seawater.

In 2005, the Environmental Protection Agency (EPA) posted baseline data for sediment contamination in estuarial coastal areas where land-based runoff settles. To obtain this data, they exposed a small crustacean, *Ampelisca abdita*, a kind of lab rat for sediment studies, to sediment samples from seven U.S. regions with marine coastlines. They targeted one hundred contaminants, including PAHs, PCBs, pesticides, and fifteen metals, a broad but not comprehensive array. If at least 80 percent of the *A. abdita* survived ten days on the sample, the sediment got a passing grade. According to this survey, Region 2, New York State, was far and away most contaminated, with 24.4 percent of its estuarial sediment proving lethal to the tiny amphipods. Region 6, Louisiana and Texas, centers of oil and chemical production, not surprisingly came in second, at a shade above 18 percent. Region 3, from New Jersey to Virginia, was next, at 9.4 percent—surprisingly low, considering substantial industry in New Jersey and Delaware and notorious pollution and fish extinctions in the Chesapeake watershed. Also surprising: coastal sediment from North Carolina down to the Florida Peninsula and back up to Mississippi tested clean as a whistle. As for my bailiwick, Region 9, contaminated sediment came in at a curiously low 7.2 percent, considering population density and intense industrial and agricultural activity in many parts of the state. Though it used a standard protocol, the EPA admitted to study limitations.

I decide to cross-reference these findings with my marine science touchstone SCCWRP. On its website, I come across a study completed in 2007. Looking for the ideal sediment "guinea pig," three staff scientists compared eleven candidate creatures, including the EPA's *A. abdita*, along with other amphipods and species of sea urchin, mussel, clam, and oyster. Their test parameters included not only lethality (acute toxicity) but sublethal impairment. Four species were trialed for "acute" testing and, lo, *A. abdita* came in last, surviving where others succumbed. In other words, if the EPA had been using the species most sensitive to California sediment, the 7.2 percent "potentially toxic" rate surely would have soared. The research team generously noted that *A. abdita* has been considered reliable in other regions, but noted a feature of the species that would seem broadly problematic: "The lower apparent sensitivity of *A. abdita* may be due to the fact that this species does not burrow in sediment, but lives in a tube-like structure and does not ingest sediment." It's actually well shielded from exposure.

So, if the reader hasn't already learned this lesson well, here it is again. Gold-standard science is hard to come by. One fact is indisputable: plastic debris, marine animals, land animals, and people all have something in common—we're oily. And that makes us all targets of insoluble, lipid-loving, man-made persistent, oily, organic pollutants.

The relevance of sediment cannot be denied. About 90 percent of ocean species spend all or part of their lives in or on bottom sediments. Let's see how polluted sediment connects with animals higher up on the food chain than microcrustaceans.

In 2008, NOAA staff teamed with scientists at the Virginia Institute of Marine Science to get a better grip on the causes of pervasive POP contamination detected in the blubber of various cetaceans, among them sperm and beaked whales, killer whales, dolphins, and narwhals. These species are known to eat deep and

BAD CHEMISTRY   245

bottom-dwelling cephalopods: octopus, squid, and cuttlefish, among others. The researchers trawled twenty-two cephalopod specimens from the western North Atlantic in waters 3,300 to 6,600 feet deep. "It was surprising to find measurable and sometimes high amounts of toxic pollutants in such a deep and remote environment," said Michael Vecchione of NOAA. The assays revealed PCBs, DDT, and brominated flame retardants, most often associated with plastics. These animals were polluted *not* in estuarial waters but near the deep-ocean seabed.

Seals and sea lions are also voracious eaters of cephalopods. At the Marine Mammal Center on San Francisco Bay, where Takada's IPW found the most highly polluted pellets, nearly 20 percent of stranded sea lions that cannot be saved have cancer, and their blubber contains high levels of persistent PCBs and pesticides. Some also have genital anomalies. Frances Gulland, who necropsied the two California whales killed by plastic debris ingestion and is medical director of the Marine Mammal Center, notes that these animals are "sentinels" of human health because much of what they eat we also eat. Chemical contamination is a strongly suspected factor in a wide range of marine mammal health problems. Autoimmune dysfunction and chemical contamination often go together in plastic-eating seabirds, marine mammals, and sea turtles. Toxic compounds are thought to be weakening their defenses against microbes and natural toxins, such as those produced by harmful algal blooms. Thyroid disorders leading to illness are common among California elephant seals. In killer whales, among the most contaminated marine mammals, calf mortalities have risen and reproductive rates languish. What's insidious about this pollution is that it doesn't kill outright. It weakens and compromises physiologic systems, creating vulnerabilities to disease, reproductive decline, and suboptimal health.

We return to the terrestrial environment, the toxic wellspring. Chemical contamination is a hydra-headed issue that penetrates nearly every corner of our lives. Plastics are one aspect, a big one, and they share a petrochemical lineage with most of the man-made chemicals we now warily encounter in our daily lives. The international chemical industry is worth a reported $3 trillion annually, and the plastics part reportedly accounts for up to 80 percent. Of the nearly 100,000 chemicals produced in the United States, 2,800 are closely watched by the EPA, being high production volume (HPV) compounds produced at a rate of at least one million pounds a year. The list does not include polymers or industrial metals. The chemicals that concern us come in two varieties. One type is a highly stable, highly mobile toxic molecule, typically a pesticide or industrial chemical, that hops a ride on plastic floaters. The others are chemicals used in the production of plastics, embedded in and later shedded from plastics into the biome. Some of the latter types have proved persistent, others not so much. But they are all bioactive. They're detectable in virtually everything we breathe, eat, and touch, including each other. It's as simple as that. And we are still in the early stages of learning how they may be affecting our health, despite studies numbering in the thousands and still being churned out, with no end in sight. It's no exaggeration to say this subject could fill a book, and, indeed, it has. And so we will pay closest attention to those chemicals that pertain to both plastics and the ocean.

Legacy pollutants such as DDT and PCBs occupy a chemical category called halogenated organic compounds (HOCs). Being highly mobile as well as durable molecules, they have been, for decades, fixtures in the marine environment. The halogens are a group of five periodic table elements: fluorine, chlorine, bromine, iodine, and astatine. The first three are germane to any discussion of plastics toxicity. These are not stand-alone elements. They are highly reactive and readily combine with other atoms and molecules. Once a bond is made, the result is a stable, "persistent" compound that degrades over decades. In biological systems, HOCs are described as "readily absorbed but not metabolized." This means they bio-accumulate, partitioning themselves into body fat and fatty organs like the liver, there to lurk indefinitely.

Fluorine is the most reactive of *all* the elements, not just the halogens. Extracted from minerals by means of electrolysis, it is then handled very gingerly. Not only is it highly toxic, its oxidizing strength is such that it eats glass. We are interested in perfluorinated compounds (PFCs), developed and deployed in the early 1950s: long-chain molecules in which hydrogen is replaced by fluorine. They are not plastics per se but they are chemical kin, often applied to synthetic polymers, and a troubling presence in the aquatic environment. Variations go by the names Teflon, Scotchgard, Gore-Tex (which *is* a polymer: an expanded fluoropolymer; it is viewed with suspicion by the chemically sensitive). These are chemicals that coat and seal, so durability is key. Starting in the 1950s, 3M and DuPont were the two producers. Both began to suspect health and environmental issues by the 1980s. PFCs were showing up in water and wildlife, and in human blood, and they did harm to test animals. But DuPont failed to inform the EPA of its findings, as required by the Toxic Substances Control Act of 1976, about which there will be more. When this came out, the EPA sued and requested that DuPont wind down PFC production. After studying its own Scotchguard division workers and finding a threefold higher rate of fatal prostate cancer, 3M reformulated Scotchgard by 2001.

In 2007, a Johns Hopkins University study of umbilical cord blood from 300 infants found traces of PFCs in all. With the intention of sourcing this pervasive human contamination, the EPA tested 116 consumer products in 2009. It found detectable traces in nanograms, or parts per billion, in waterproofed apparel, upholstery, home fabrics, and carpeting; in carpet care products and floor wax and wax remover; in stone and tile sealants; and, interestingly, in "nonwoven medical garments." Some nail polish brands contain perfluorinated compounds. According to a 2007 item carried by Foodproductiondaily.com, which, as a trade "publication" will often provide enhanced information about industry practices, "PFOA (perfluorooctanoic acid, a common PFC) has been employed in the manufacture of grease-resistant packaging for candy, pizza, microwave popcorn, and hundreds of other foods," including butter and fast-food containers and wrappers. It's the Chemical Age version of waxed paper. Its ubiquity would be concealed beneath the protective coating of *trade secret*. Until health concerns emerged, PFCs had been used widely in electronics manufacturing, both as a surface sealant for circuit boards and as a mist suppressant during fabrication.

Waterproofed products for decades have been impregnated with PFCs, among them shoes and boots, luggage, and camping and sporting equipment, including backpacks. Because this chemical class has proved both persistent and toxic, efforts to stop it at its source have been concerted and international. PFCs have been detected in fish, loons, sick sea otters, and endangered loggerhead sea turtles, whose blood samples showed markers of liver damage and compromised autoimmune function in a 2010 study. A marine mammal health assessment project found that PFC levels in dolphin tissue on average doubled between 2003 and 2005. DuPont itself admitted dumping tons of PFCs into water systems for years. From there, it went into the food chain. In lab trials, rodents and monkeys have been administered PFCs

in varying doses and these were the results: reduced size and
greater mortality in offspring; slowed physical development;
liver, testicular, and pancreatic cancers; thyroid dysfunction; and
altered lipid metabolism, meaning they may promote obesity.
These happen to be among the "new" health epidemics. A 2010
Forbes magazine report on most-prescribed medicines put levo-
thyroxine, the standard treatment for people with underactive
thyroid function, at number four, with 66 million prescriptions
per year. Huey, our rug-hugging cat, needs his twice-daily thyroid pill, making him part of a growing number of pets with
chronic disease. Research conducted at Indiana University found
dogs and especially cats to be loaded with persistent synthetic
chemicals. Pet health trends include cancer, obesity, and diabetes.

Human studies conducted by U.S. and European universities
have found correlations between higher PFC levels in blood
serum and early menopause, thyroid dysfunction, and ADHD,
as well as "subfecundity," meaning fertility problems. A stronger
link between PFCs and human injury occurred at a West Virginia
DuPont plant where several female workers were reported to
have given birth to babies with facial deformities similar to those
seen in a cohort of exposed rats. Occupational exposures that
cause harm pose interesting questions. While they're worst-case
scenarios unlikely to be experienced by the typical consumer,
they offer the only de facto human test results we have—proof,
in many cases, of "potential" human risks suggested by animal
studies. Such grim outcomes also support the argument that
large-scale production of any toxic substance is a devil's deal.
Surely we can find green chemistry equivalents or simply do
without chemicals that poison and persist. The EPA deems PFCs
to be "potential" health risks, and "ordered" a voluntary phasing
out of their use in consumer products in the early 2000s. While

not so widely used as they once were, PFCs will haunt us in the persistent way of POPs.

I've come across a few reports of PFCs adsorbed by debris plastics, but this chemical family is not targeted with the same intensity as legacy POPs, which populate the next category of organic halogens: *chlorinated* organic compounds. These are the granddaddies of man-made toxics that seemed such a good idea at first. Pesticides like DDT and those tenacious industrial molecules, PCBs, still linger like unwanted guests. Their relevance to plastics toxicity is a matter of dangerous affinity. They're non-soluble, lipophilic, and highly mobile, so they readily fly and flow into the oceans. Some, as we've seen, bind with sludgy seabed sediments and contaminate bottom-dwelling creatures. Others float until encountering and latching on to something oily, such as krill or a plastic floater. Though these legacy POPs are measurably waning, a 2007 Flemish study showed that local eels caught near industrial areas still harbored levels of PCBs worthy of a warning to Belgian eel lovers to lay off the wild-caught variety until further notice.

What is most distressing is the impact of these POPs on remote human populations. Some are contaminated at levels equivalent to occupational exposure. Indigenous enclaves in Northwest Greenland and Canada are among the most polluted on the planet. In 1991, the Canadian government's Indian and Northern Affairs Canada agency set up the Northern Contaminants Program. Working with Inuit hunters and trappers during the hunting season, scientists obtained tissue and organ samples from fish and game staples, with ringed seals providing an especially rich trove of biomonitoring data. The most recent report from 2005 found that PCB and DDT levels had fallen more than 50 percent since peak readings in 1975—shortly after the U.S. DDT ban and shortly before its PCBs ban went into effect. But newer POPs were on the rise, and the studied creatures also harbored elevated levels of toxic metals, primarily methylmercury and cadmium. The levels were sublethal, as toxicologists would say, but there are disturbing health trends, the most conspicuous being radically skewed girl-boy birth rates. The most contaminated people ever biomonitored live in northwest Greenland. There, two girls are born for every boy. The mother's lipid PCB level correlates with her likelihood of delivering a girl. Moreover, the Danish scientists who studied this community found that the babies tended to be born preterm and small, which often portends developmental and neurological problems. This is a sign of endocrine disruption, of genetic blueprints being altered by subverted hormonal messaging during gestation.

How do people who live as far as one can get from centers of industry become more contaminated by industrial chemicals than, say, a Brooklynite? For the answer we return to the oceans. Inuit people, those few who hew to traditional ways, are apex predators in a food web that starts with plankton. They are as close as people get to being "marine organisms," subsisting mainly on seal, whale, and fish, and lesser amounts of caribou and reindeer. These dauntless souls occupy a top rung in the marine food web and thus unjustly illustrate the principles of bioaccumulation and biomagnification. These phenomena are really at the heart of what's so pernicious about the way we're contaminating the oceans with chemicals and the plastics that so effectively concentrate and transport them, often into unsuspecting mouths and beaks.

Here is how bioaccumulation and biomagnification operate in the oceans:

On the first level we find "primary producers"—phytoplankton (microscopic plants) and algae, which subsist on sunlight and photosynthesis. You may have heard that alternative-energy companies and even a few oil giants are developing fuel from algae, which tells you that algae will attract oily substances, including POPs. In polar regions, both north and south, studies from the early 2000s showed that airborne POPs—carried by circulating wind currents similar to those in the oceans—condense into sea ice, then melt into arctic waters during "summer" months. This accounts for legacy POP contamination in these regions persisting at a higher levels than in most parts of the planet. The spring melt coincides with annual phytoplankton blooms, meaning plankton get marinated in the stuff.

Each layer in the food chain is called a trophic level. Localized versions of the food chain are called food webs. The second and third layers, above microorganisms, include filter feeders, like salps, fish larvae, jellyfish, lanternfish, and the all-important krill. These creatures would be bioaccumulating toxics from surrounding waters while reaping the multiplier effect called biomagnification. Biomagnification occurs when an animal's dietary intake is polluted, making for continuous dosing and accretion of toxic residues. Top-level predators in polar regions, mainly whales and seals, have thick blubber layers teeming with stored POPs, and also toxic metals. And so it is that the Inuit are polluted by their "natural," traditional diet, a fact they are aware of and unhappy about. In her 2010 book, *Empire of Ice: Encounters in a Changing Landscape*, Gretel Ehrlich, a writer who is drawn to cold places and the people who live in them, quotes an indigenous Greenlander: "We're talkin' social justice issues, we're talkin' heavy metals, radioactivity, mercury, soot, and POPs. We're carrying mercury from your coal plants, eating your endocrine disruptors, drinking your soot. . . . And we are the last traditional ice age

hunting people in the world." He knows more ecotoxicology
than the typical Brooklynite.

Here I would like to make an unequivocal statement. I believe
trophic-level diagrams of the food chain are inaccurate and require revision. They misrepresent reality by ignoring plastic at
every level of the food chain. Plastic is eaten on each trophic level,
despite being non-nutritive and possibly toxic. It exists even at the
sub-planktonic layer occupied by microscopic bacteria, which,
we have learned, "eat"—that is, partially biodegrade—plastics,
and on up to apex predators like albatross and whales, as we have
seen. Plastics could even be considered, in a sense, predators,
given the deadly nature of "ghost fishing" and entanglements of
marine turtles, mammals, pinnipeds, and cetaceans. Though plastic is not a living organism, it acts like one and has the impact of
one and should be taken into account in characterizations of the
ocean biome. What is most shameful in this more realistic modern
scenario is that plastic, in a sense, is man's surrogate, swimming
with the fishes and doing harm.

We've also seen how plastics sneaked into the terrestrial food
chain, proving lethal to camels in Dubai and cows in India. Now
I've learned of an industry that purposefully injects plastics into
the human food web. Since the 1970s, some feed purveyors have
blended plastic pellets into "concentrated" feed, the type used to
"finish" beef cattle. The practice was well studied in animal hus-
bandry programs at land grant universities, and proved a form
of "roughage" that would enhance nutrient uptake and growth.
One has to wonder what becomes of the plasticized manure, if
it is composted and added to agricultural soils. There's a conspicuous lack of readily available information about the practice.
But I have it on good authority. My coauthor's husband worked
at a Kansas feed and grain outlet in the 1970s and loaded many

pickups with sacks of plasticized feed. As an ongoing practice it is confirmed by inference in dozens of online extension service instructional sheets about organic ranching. To produce organic beef, you need to follow several cardinal rules, and one of them is not to feed the cattle plastic roughage.

DDT is finally fading from the scene, despite its use in some equatorial countries where its toxic risks are outweighed by the threat of malaria. In several African countries where its use persists, trends to low-birth-weight babies have been observed. But more widely and recently used PCBs still linger and still show up in human breast milk. PCBs are what our Japanese colleagues mostly detected in plastic pellets, both new and feral, in and around Tokyo Bay, and around the world in the International Pellet Watch studies. We have learned the acidic digestion process releases toxics from plastics into biological systems. It cannot be said that plastic-eating seabirds are only toxified by poisoned plastic: their fish diet is also contaminated. But these "poison pills" have been shown to augment chemical burdens and add to immune system dysfunction, which appears to be a primary effect of PCBs on seabirds.

The standard line on PCBs is that they were industrial chemicals commonly used as insulators and flame retardants in electrical transformers (some of which are still in use). That hardly seems to account for what appears to be near-universal body contamination by these molecules. A report from early 2011 brings the unpleasant news that today's children may be exposed via PCB-laden caulking in older buildings, including schools. Connecticut's Department of Public Safety offers an enhanced list of former PCB applications. It includes adhesives, asphalt roofing materials, carbonless copy paper (blamed by some for causing "multiple chemical sensitivity syndrome"), caulking,

compressor oil, de-dusting agents, dyes, fluorescent light bal-
lasts, inks, lubricants, paints, wood floor sealant, pesticides, plasticizers, rubberizers, space heaters, tar paper, and wax extenders
used for toilet installation. While no products manufactured
since 1978 would (legally) contain PCBs, older PCB-containing
products are still very much among us, making them legacy contaminants in more ways than one. Indeed, older homes, thrift
shops, and garage sales will ensure their presence among us into
the foreseeable future.

How exactly do these seemingly embedded toxics get into
the bodies of humans and other living creatures? Toxicologists
note three basic routes of contamination: ingestion, inhalation,
and absorption through our largest organ, the skin. Ingestion isn't
always about food. We also swallow as well as inhale contaminated dust. Studies show contaminated household dust to be a
significant chemical vector. As to inhalation, the brand-new
shower curtain offers a peerless example. Who doesn't know that
potent "plasticky" smell that somehow we've come to associate
with "new" and "clean"? In 2008, Canadian researchers decided
to quantify fumes gassing off five new PVC shower curtains
bought at several big-box outlets. Over 28 days they logged, to
their amazement and mine, 108 chemicals, mostly volatile organic
compounds (VOCs)—noxious solvents such as benzene and
toluene—as well as phthalates. Some concentrations exceeded
Occupational Safety and Health Administration (OSHA) safety
standards for several days before dissipating to "normal" levels.
Prolonged exposure to many of these chemicals is associated with
respiratory irritation, headaches, nausea, and potential harm to
the liver, the kidneys, and the central nervous system. They can
also cause cancer. Alone and over time, a vinyl shower curtain
could be considered an acceptable risk. But in concert with

dozens of other well-studied household contaminants that we eat, breathe, and touch, the risk is amplified.

Many things that we imagine to be stable, solid, and inert are not. Unseen by the human eye, molecules lead their own little lives. Over time, a PVC window frame, for instance, will take a beating from the sun's ultraviolet rays, or you will see little cracks form in a hard plastic tumbler (polycarbonate) after repeated hot dishwasher washings. Over time, oxidation occurs, cracks and pits form, polymers (that is, molecule chains) weaken. This liberates unbound, single-molecule additives. According to the plastics text *Industrial Plastics: Theory and Applications* by Terry Richardson and Erik Lokensgard, these are the "major categories" of additives: antioxidants (both primary, used during the production process, and secondary, added to the final product), antistatic agents, colorants, coupling agents, curing agents, flame retardants, foaming/blowing agents, heat stabilizers, impact modifiers (strengtheners), lubricants, plasticizers, preservatives, processing aids, UV stabilizers, and anti-microbials. That's a lot of chemicals, and each category is loaded with options, many of which come from toxic categories: phenols, glycols, heavy metals, solvents, and biocides. Before 1993, the toxic metals lead, cadmium, mercury, and hexavalent chromium (the *Erin Brockovich* toxicant) were allowable inorganic pigments. The concern when they were banned for this use in the 1990s had less to do with direct human contact than groundwater contamination from landfill leachates. Plastic food packaging applications require FDA safety approval. But this doesn't always mean a thorough lab workup or even agency access to the company's chemical recipe for patent-protected materials. Regulatory policy takes into account harm to a company's bottom line and seemingly trusts the marketplace to sort things out—except when a chemical causes

bodies to drop or becomes a political football. Both are usually
too late, especially if the harmful agent is the persistent variety.
What's often the case in these scenarios is that the "sentinels" of
harm are non-human creatures, typically aquatic creatures, with
lab rats soon to follow.

THE THIRD HALOGEN IS BROMINE. One of its forms has already
been mentioned: polybrominated diphenyl ethers or PBDEs.
Bromine, derived from salt, is another highly reactive element
that requires chemical partnering. Brominated flame retardants
(BFRs) work on the premise that heat causes their molecules to
break apart and release bromine that snuffs the flame. They come
in myriad formulations of varying molecular lengths—with the
shorter ones being the more toxic. Structurally, they are similar
to PCBs. Several states—including Washington, Maine, Michigan, and California—and the E.U. have already banned or restricted use of all or several BFRs, convinced by mounting evidence
that they harm wildlife and human health. European countries
never really took to them. Just how widely they've been used in
plastics is anyone's guess. This would be proprietary information.
A reasonable assumption would be that their use was wide and
deep, but is less so now as flame retardants reported to be less
toxic come to market. Because they're exposed to heat, plastic
casings for electronics like cell phones, computers, and TVs contain flame retardants as a matter of course. So do plastics and
synthetic fibers in cars and commercial jets, as well as fleece blankets and foam mattresses, one brand of which was recently recalled by the Consumer Product Safety Commission for lacking
flame-retardant chemicals. Safety-minded Americans embraced
brominated flame retardants with a vengeance in the 1970s, to

replace banned PCBs. In a classic case of good intentions gone weird, the state of California installed the nation's highest flame-retardant standards for children's products and household furniture. As a result, California households and bodies are the most BFR-contaminated of any ever tested. In an abrupt one-eighty, the California legislature voted in 2003 to ban two forms of PBDEs. The chemical industry managed to extend the phaseout deadline two years, from 2006 to 2008. Over time, Californians' flame-retardant body burdens should decline, but damage may already have been done. Studies have correlated high infant exposure to PBDEs with diminished intellectual capacity. Like PCBs, BFRs generate highly toxic dioxins when burned. This has led to one of the odder ironies yet: firefighters lobbying against flame retardants. While a PBDE-impregnated cushion might extinguish a dropped cigarette, PBDEs are no match for a major fire. Then the problem becomes toxic smoke, which is more likely to cause injury or death than flames. Nowadays, house fires put firefighters "in defensive mode," according to a Missouri fire chief.

By the late 1990s, brominated flame retardants were being compounded with 2.5 million tons of synthetic polymers annually. It was around then that safety questions began to arise, after two decades of swiftly escalating use. BFRs were also turning up in monitored wildlife—fish and bird eggs, harbor seals in San Francisco, and killer whales near Seattle—and in sediment and sewage sludge. PBDE levels in Great Lakes walleye and lake trout rose exponentially from 1980 to 2000, doubling every three to four years. PBDEs were also turning up in food, especially fatty meats and salmon, in human breast milk, and in household dust. People who eat lots of beef and chicken have higher BFR levels. A 2002 Swedish study pronounced PBDE health effects as strong as or possibly stronger than those of PCBs. With several states banning PBDEs and electronic and mattress companies forsaking them, and even Walmart phasing out products containing them, why hadn't the EPA ever taken action? Something is very wrong with this picture.

PCBs were banned along with a group of pesticides in the late 1970s, shortly after passage of the Toxic Substances Control Act (TSCA) of 1976. PBDEs soon filled the flame-retardant breach, but under TSCA rules they were not required to be proved safe, due to an astonishing loophole. The act required chemical producers to provide data in order to obtain registration, but required no toxicity testing. Safety testing would only be triggered by strong evidence of harm to human health—meaning people sickened by the chemical—provided by the EPA. Only then was the EPA permitted to order a company to conduct safety trials, which is something like throwing Br'er Rabbit into the briar patch. The TSCA is so industry friendly that even asbestos, a known carcinogen, could not be banned under its guidelines and remains, technically, approved. Meanwhile, thirty other countries have banned its use. The EPA's efforts to ban asbestos, or any other toxic chemical, ended in 1991, affirming the TCSA's impotence. Subsequent phaseouts of suspect chemicals have resulted from voluntary actions requested by the EPA. Individual states, however, are allowed to ban chemicals, and, as we've seen, they've exercised this prerogative. California led the way with Proposition 65, requiring that businesses disclose through labeling a product's ingredients known to cause cancer, birth defects, or other reproductive harm. The American Chemistry Council (ACC) is not keen on this state-by-state "patchwork" approach, preferring instead to guide congressional efforts to reform the TCSA in ways that would disempower states. The ACC is spread thin enough fighting local and state efforts to ban plastic shopping bags.

In April 2011, a new Senate bill arrived, the Safe Chemicals Act of 2011, which calls for, among other measures, assessment of tens of thousands of chemicals that escaped scrutiny under the toothless 1976 law. The ACC, while professing support for TSCA reform, expressed disappointment in the proposed legislation: "Unfortunately, it appears many of our concerns have not been addressed . . . and the bill introduced today could put American innovation and jobs at risk." In other words, if the new law requires light to be shed on toxics now in common use and toxicity testing for chemical "mixtures," and if it unmasks protected "proprietary" ingredients in everyday materials and products, business as usual ends. I would suggest the ACC look on the bright side: lots of jobs and innovation will be needed to create safe alternatives.

After years of holding back, a growing number of mainstream health organizations have joined eco-advocacy groups in support of TSCA reform. These include the American Medical Association, the American Academy of Pediatrics, the Endocrine Society, and the Autism Society. All have gone on record with concerns about suspected links between chemical exposures, also called environmental exposures, and a burgeoning array of chronic diseases and disorders. These so-called new morbidities include obesity, type 2 diabetes, autism, ADHD, asthma, thyroid disorders, and male infertility. Of special concern are exposures that cross placental barriers and appear to alter gene expression in the developing fetus. Thus far, the marketplace has, by default, proven a more effective regulator than any agency. When a chemical like bisphenol-A (BPA) gets bad press, manufacturers scramble to get BPA-free products on store shelves, an effort that may, in fact, be wasted. Recent tests by University of Texas biology professor George Bittner showed more endocrine disruption in some BPA-free products than the originals.

While the United States abides by a "proof of harm" rule, Europe applies the precautionary principle. That means their risk analysis favors human health over the economic health of chemical producers. As a consequence, Europeans have measurably lighter chemical body burdens than do Americans, especially in the category of brominated flame retardants.

As first-generation BFRs are phased out, newer ones come online, including nano formulations. Still in the picture is Tris, a chlorinated flame retardant that made headlines in the 1970s, when it was banned from children's sleepwear, but not from other applications. Animal studies with Tris were showing liver and kidney toxicity as well as neurological effects, including cognitive impairment. The fact that phased-out BFRs still register strongly in annual human sampling conducted by the Centers for Disease Control and Prevention is mystifying even to the EPA, which has theorized shedding from household sources—such as older carpet, furniture, and electronics—as well as leaching from landfills into water supplies. What now seems very clear is that their use was profligate, because the scope of contamination is epic. Our plastic foams, solids, and fibers—our furniture, carpets, drapes, and car seats, most made of polymer fibers—all contain chemical flame retardants. Where do you least need flame retardants? At the polar regions, perhaps? Just as those remote ecosystems began to see glimmers of relief from legacy POPs, BFRs arrived on the scene.

The most concerning effects of these chemicals are on the thyroid. During fetal development, the thyroid gland carries out much of the signaling that determines proper aligning of cells in the developing fetus. This includes brain development. If the mother's thyroid function is severely depressed and remains untreated—which is rare in developed countries—cretinism can result. If it's slightly depressed, the child may still lose intellectual potential. When PBDEs began to show their harmful colors, new and "improved" brominated flame retardants with longer molecular chains took their place. Within years they were showing up in the same monitored wildlife populations, and the chains were breaking down into shorter, more toxic units. The European Union banned the whole lot by 2004, and the Stockholm Convention put them on ban lists, along with PFOs, the Teflon-type chemicals.

In September 2010, a group of activist-scientists met in San Antonio, Texas, and formulated what is known as the San Antonio Statement on Brominated and Chlorinated Flame Retardants, documenting, as the Green Science Policy Institute put it, "the health hazards and lack of proven fire safety benefits from the use of brominated and chlorinated flame retardant chemicals in home furniture, baby products, and other consumer products." Since then, the document has gained more than two hundred signatories, most of them prominent scientists and doctors, from thirty countries. These include European countries where BFRs are already banned. Why should they bother? Because our flame retardants affect their ecosystems, and the global food chain. These scientists list among their concerns not only the diseases and disorders already noted but no less than the erosion of human intelligence. Recall that California may be the most BFR-saturated place on earth. Now a Tufts economist has calculated a probable loss from BFR-caused IQ impairment in California at more than $50 billion. Literally dumbed down, a society loses talent, productivity, competitiveness, and possibly the ability to organize politically to combat its own decline.

One of the signatories is Dr. Susan Shaw, founder of the non-profit Marine Environmental Research Institute based in coastal Maine. Shaw maintains the Seals as Sentinels project, which monitors the local seal population for evidence of chemical contamination and illness. By measuring the chemical burdens of stranded seals, she's discovered that BFRs not only persist but bioaccumulate, just like PCBs and DDT. Maine is not highly industrialized, but BFRs roam the planet. The seals ingest many of the same creatures we eat—haddock, cod, and crustaceans—which we can deduce to be contaminated. Of particular concern to Shaw are mass seal die-offs that occur with increasing regularity. Shaw considers it probable the animals' chemical burdens have lowered their natural immunity. She wrote in a piece for the Explorers Club: "Mounting data suggest that by continuing to pollute our seas, we may be irreversibly poisoning marine life and ourselves."

In spring 2011, Dow, the largest U.S. chemical company, announced the invention and market readiness of a new polymeric brominated flame retardant. They are calling it Polymeric FR, and the first licensee is Chemtura, a Michigan chemical company that made PBDEs and filed for bankruptcy protection in 2010 as red flags were flying about its product. Now Chemtura is back in the game. According to Dow's press release, this "next generation" material is designed for use in "both extruded and polystyrene and expanded polystyrene foam insulation applications globally." It's a product of "Dow's continuing search for more sustainable products," that is, products that don't run the threat of being banned. Polymeric FR has, the company says, been thoroughly safety tested, by Dow, and found to be stable but not bioaccumulative or toxic. Under current law, if Dow's testing revealed toxicity, and if it was discovered that Dow withheld that information, the company could be fined. (This policy has discouraged a lot of industry testing.) The best we can do is hope Dow is leveling with the EPA and its customers. If this "stable" molecule winds up in amniotic fluid and mothers' breast milk—and in sick sea otter blood samples—we'll have been duped yet again.

With the persistent, bioaccumulative halogen toxics— brominated flame retardants, the fluorinated coatings, and the chlorinated lubricants and pesticides—the pattern has been this: industry has a good run with them, while dribs and drabs of troubling evidence keep mounting. There comes a tipping point when a chemical's purported benefits cannot justify what appear to be harmful unintended consequences. And when that point comes, the chemical industry unveils a replacement, or several. Meantime, because law does not require disclosure of ingredients in material products, and because the ingredients in industries' products are so often hidden behind "trade secrets" law, we don't really know what we're being exposed to.

This may be true as well for those notorious plastic chemicals that have turned so many Americans into amateur toxicologists. These would be, of course, BPA and phthalates, both very hot topics over the past half decade. Volumes could be written about just these two chemical villains, and they have been, so our consideration will take a slightly different tack. Tens of millions of research dollars have been spent trying to prove their safety or harmfulness, and still the science is unsettled, at least in official eyes. BPA is a mildly estrogenic synthetic phenol, produced as a powder, then industrially reacted with phosgene, a poisonous gas used in World War I, to make hard, clear polycarbonate plastic. Until recently, polycarbonate plastic was the standard material used to make baby bottles, five-gallon watercooler bottles, and reusable water bottles before the marketplace took sharp corrective action. It's still the preferred material for eyeglass lenses, goggles, helmets, and appliance casings, among other things. BPA is also in the epoxy lining for canned foods and beverages, and it can be detected in the urine of 95 percent of Americans. Phthalates are a family of synthetic esters that soften hard vinyl to make things like chew toys, air mattresses, shower curtains, vinyl flooring, and Naugahyde, and they are "fragrances" in personal products. Phthalates are detected in more than 95 percent of Americans. In 2008, the most bioactive phthalate, DEHP, was banned from children's toys, not by the EPA, but by a congressional bill that was signed into law by President George W. Bush.

Phthalates and BPA are different from the halogenated organics. They don't persist and they don't bioaccumulate. They have been detected in low concentrations in waterways and sediments near sewage plants and landfills, but not in the tissues of the ocean's apex predators. They are metabolized by the liver and excreted, though they are now known to linger longer in the human system than once believed. The fact that they are metabolized and the findings that Americans are so widely contaminated by them suggest constant exposure. Could it be that phthalates and perhaps even BPA are found in plastic materials not normally associated with these chemicals—that is, plastics other than PVC and polycarbonate? Could that be why Americans verge on being ambulatory superfund sites?

I wasn't seeking the answer to this question when, in 2005, Algalita undertook a study of chemical contamination in plastic debris and loose nurdles recovered on land. The effort was included in the study commissioned by California's State Water Resources Control Board, and it entailed collecting samples both from Los Angeles riverbeds and outside a local plastic pellet distributor. The focus was on POPs. Hideshige Tadaka and his colleagues had already learned plastic pellets adsorb PCBs and DDT in the marine environment. We wanted to see if this type of contamination was occurring before plastic pellets and fragments reached the ocean. So off we went, Algalita staff biologists Ann Zellers and Gwen Lattin and I, armed with tweezers and metal collection bowls. We harvested our samples, then labeled and sent them to a state certified lab for the complex extraction and analysis process.

The results were very different from those obtained by Takada and his team: no PCBs or DDE (a breakdown product of DDT), but every sample contained traces of toxic PAHs, likely from L.A. air pollution. They also, almost universally, included phthalates. Phthalates are not the sort of chemical to be drifting around in the environment, attaching to stray plastic pieces. So we had to conclude that they were ingredients in the samples themselves. This would have made sense if we'd been gathering soft vinyl rubber bath toys or peel-and-stick floor tiles, but we weren't. All of our samples were polyethylene and polypropylene. The phthalates must have been compounded into the PE and PP samples, not adsorbed from the environment. If this is a regular practice, industry has made chumps of us all, and done so by exploiting the legal cover offered by trade secret protection.

Is BPA also more widely used than we suspect, perhaps in other types of hard plastics? Results from International Pellet Watch analyses suggest yes. Both polyethylene and polypropylene pellets and fragments harbored levels of BPA comparable to and in some cases exceeding those of legacy industrial chemicals and pesticides, ranging between 1 and 1000 parts per billion. A key issue with this chemical is the health effects some researchers have seen in animals administered doses at very low levels. In laboratory experiments, BPA at parts per trillion have suppressed testosterone production, and amounts comparable to those in the gyre studies have caused "adverse health effects" in mice. Frederick S. vom Saal and Claude Hughes, premiere BPA researchers and opponents, wrote in a 2005 piece for *Environmental Health Perspectives:* "Because there is evidence that BPA is rapidly metabolized . . . these findings [that 95 percent of Americans have detectable BPA in urine samples] suggest that human exposure to significant amounts of BPA must be continuous and via multiple sources." This begs two questions: are polycarbonate plastics far more ubiquitous than we realize? The ACC's resin production statistics don't even list polycarbonate, presumably because its production rate is paltry compared to PE, PP, PVC, and the others. The second question would be: are 95 percent of Americans eating that much canned food and drinking that many canned beverages on a daily basis? It begins to seem likelier that BPA is a hidden presence in a host of other plastic types to which we live in close proximity. When was the last time you saw a label on a car steering wheel, cell phone, or laptop computer telling you its chemical ingredients?

So, we turn again to the International Pellet Watch project and the results from analysis of our gyre samples. While our samples were among the more pristine when it came to "sorbed" pollutants, legacy chemicals such at DDT and PCBs, they earned the dubious distinction of being the most contaminated in two categories, no small feat in a field of fifty-one different sites, many of them urban/industrial. Of the polyethylene samples from all locations, ours contained the highest levels of nonylphenols, a chemical cousin to BPA most commonly used as a surfactant in detergents and pesticides, but also as a plasticizer and antioxidant in plastics. In the late 1980s, a study involving breast cancer cells at Tufts University went haywire when the cells wildly proliferated instead of shrinking. It turned out the manufacturer had added estrogenic nonylphenol to the plastic used to make lab ware. Are gyre zooplankton skewing feminine? Might toxic marine bacteria be proliferating? Are male lanternfish experiencing a loss of mojo, and might this entail possible reproductive consequences? We do not know.

Among all polypropylene pellet and fragment samples, ours from the gyre were by far the most contaminated with decabrominated diphenyl ether (BDE209), a next-gen, longer molecule brominated flame retardant now shown to break into smaller, more toxic segments. Toxic flame-retardant plastics in the middle of the most massive body of water on earth. The word does not exist to express the irony plus wrongness of this finding. We have just seen the types of health effects this chemical category is capable of delivering.

A summary of the International Pellet Watch project to date is titled "Global distribution of organic micropollutants in marine plastics." Takada is lead author and I am one of thirteen coauthors. But it was he who elected to conclude the report with the following point: "Even in the open ocean and the remote coast, high concentrations of additive-derived chemicals, such as nonylphenol and decabrominated diphenyl ether (BDE209), were found. . . . In the open oceans and remote coast, ecological risk associated with plastic additives could be more serious than chemicals sorbed from seawater."

In other words, the plastic material is itself a toxic Trojan horse, not so bad to look at but riddled with unseen chemicals that may pose greater risk to marine—and terrestrial—biota than the dreaded persistent organic pollutants. It will be up to the next generation of marine scientists to investigate this potential threat.

CHAPTER 14

DEBRIS FORENSICS

TAMING PLASTIC POLLUTION may prove more difficult than banning toxic chemicals, and devising means to both ends poses many similar problems. With polluting plastics, their destinations are as multifarious as their sources, and the two often bear little or no connection to each other. But the more we learn about these sources, the better the odds for halting the plastic torrent. We begin with a story.

Seal Beach, just south of Long Beach, never had many seals. California *sea lions* mostly, and a few harbor seals, basked on its sloping sands before coastal development pushed them out. But it's still a fine place to surf, and it happens to be home to an old friend, Judy Naimi-Yazdi. She knows this shoreline as intimately as anyone. On regular visits, this longtime member of the Surfrider Foundation collects litter and maintains pollution vigilance. One day she sees an unfamiliar sight: hundreds of blue plastic spirals strewn across the sand, looking like hair curlers dropped from the sky. Because I was, at the time, head of the Blue Water Task Force, the Surfrider water testing program, news of the mysterious spirals comes to my attention. This was in the early 1990s, before my first trip through the Garbage Patch, but plastic beach litter was already an established scourge. I contact a friend, a sales rep in the recycled plastics industry. Even he's mystified. He takes a handful of the spirals and promises to make inquiries.

Debris forensics is using what we know or can learn about plastic trash to trace it to its sources and stop it there. The term aptly evokes crime scene investigations because plastic trash in the oceans is a violation of international law, as the reader well knows by now: MARPOL Annex V, in force since 1988. Federal law also prohibits dumping plastics in the U.S. exclusive economic zone—ocean waters within two hundred miles of U.S. shorelines. The U.S. Coast Guard requires vessels longer than twenty-six feet, including ORV *Alguita*, to conspicuously post a sign that lists all materials legally barred from ocean waters. The crime metaphor becomes literal in light of well-documented harm plastic debris does to marine species. Sylvia Earle, the noted oceanographer known as "Her Deepness," captured the balefulness of plastics when she wrote in her book *Sea Change* that even though plastics lack their own "detectable aroma . . . [they] bring to the sea the smell of death."

Location counts when it comes to plastic debris, with urban shoreline litter being a good deal more traceable than mid-ocean flotsam. Once plastic trash goes pelagic, it's fled the scene of the crime and the trail goes cold. Even the most advanced forensic methods won't get you far. A technique for dating these pieces would, at least, illuminate the mystery of how long plastics take to degrade to nothing in the marine environment. One theory holds that small bits become rounder as they age. While this may be a plausible hypothesis, it's yet to be subjected to scientific scrutiny and is thus an impermissible forensics tool. We may simply have to accept that fragmented plastics are cold cases, and move on.

While he's still working on a way to hydrocarbon-date broken-down plastics, Anthony Andrady has charted the agents and stages of plastic "degradation." His work has borne fruit if not yet all the answers. He has decoded the agents and stages of plastic decay, but there's no way to predict how long these processes will take, given myriad variables in natural settings. The worst enemies of plastic are known: heat, sunlight, and mechanical abrasion. To lesser extents, air, water, and biological organisms abet degradation. Polymers are chemically bonded microscopic fibers. Exposed to the elements, the bonds become stiff and arthritic and then brittle, like a plastic bucket forgotten in a corner of the garden that crumbles in your hands. The last stage of degradation is mineralization, when the polymer molecule collapses in an invisible puff of carbon dioxide, water, and trace minerals. But you don't have to be a polymer scientist to know a thin plastic shopping bag will degrade faster than a PVC lawn chair.

One of Andrady's experiments, conducted with J. E. Pegram, compared the degradation rates of typical marine litter samples in different environmental conditions. He suspended sections of nylon trawl webbing and polyethylene and polystyrene strapping tape in seawater and placed equivalent samples nearby on land, in two very different locations: Biscayne Bay in Florida and Puget Sound in Washington State. At the end of the study, the materials would be tested for loss of "mechanical integrity." A year later they found that the samples in seawater at both sites were essentially unchanged. But after only six months the sam-

ples on land exhibited "a severe loss in mechanical integrity."
The darker-colored plastic pieces were heat "sinks"—measuring
up to 86 degrees Fahrenheit hotter than ambient air. His conclusion: the ocean is a sort of Shangri-La for plastics. Degradation
there occurs in slow motion, with a suite of factors making this
so: water is cooler than land; saltwater is less oxidizing than air;
and algae "fouls" marine plastics—that is, it coats and shields it
from the sun's photo-degrading rays.

In light of this data, our gyre samples begin to look less like
ocean-based plastics that have degraded over time—though some
undoubtedly have—and more like plastics that baked on a sunny
beach or stretch of blacktop in a prior terrestrial life. In fact, a
New Zealand researcher found that after only a few weeks on a
hot beach, certain types of plastics, such as "crisp" bags, were
already thin and brittle. This finding seems to support the notion
that much of plastic flotsam, at least the tiny bits, comes from
land, pre-stressed. It also reinforces the importance of preventing
plastic garbage from going into the oceans, where it lasts longer
and wreaks more ecosystem damage. The ugly eyesore on land
becomes a villainous menace in the sea.

We've already had a glance at the systems environmental agencies use in fashioning pollution remedies. Pollutants are divided
into two categories: point source—traceable to a specific entity,
such as an illegal dump or factory—and *non*-point source—
untraceable to a particular polluter, like roadside trash or plastic
specks in the sand. These rules rarely apply to the mid-ocean,
where ship dumping is likely to include consumer plastics as
well as fishing gear. It's hard to see the end to the debate of land-
versus ocean-sourced debris, and a catastrophe like the 2011 Japan
tsunami tells us that proportions will vary. Andrady suggests
plastic flotsam comes more or less equally from land- and ocean-

based sources. It's as reasonable an estimate as I've heard, with one caveat: plastics lubricate globalization and plastics pollution results from globalization—especially in regions lacking adequate disposal systems. The rivers and shorelines of many developing countries are choked with plastic bags, bottles, wrappers, and containers that will make their way to the oceans. By all rights, the makers and purveyors of these plastics should be responsible, but this is not yet the world we live in.

But everyone can agree on a few things. The ocean-based sources of plastic trash are, according to the UN Environment Programme, merchant ships, ferries, passenger ships, fishing vessels, military and research vessels, pleasure craft, offshore oil and gas platforms, and aquaculture farms. Another category I'd add is shipping containers lost at sea, whose contents have provided opportunities for the study of ocean currents. The main land-based sources of marine litter are landfills located on or near the coast, rivers, and waterways; treated and sometimes untreated sewage; industrial facilities; and "tourism." Another category is, of course, natural disasters: monsoons, hurricanes, tsunamis, and earthquakes. Any of these can and usually do release massive amounts of debris into the oceans. Debris in deep-ocean garbage patches, and on remote collector beaches in Alaska and Hawaii, are similar: heavy on fishing gear, Asian products and containers, and plastic flakes.

Tracking plastic trash sources is further confounded by the extreme mobility of the material. Plastic litter comes and goes. It's swept out to sea by wind and tides and tossed back onshore by currents and waves; it sinks, swims, floats, and flies. It gets passed up the food chain by migratory creatures. The famous 1992 container spill of twenty-eight thousand Friendly Floatee bath toys enabled Curtis Ebbesmeyer and James Ingraham to test and refine their OSCURS model of global ocean currents because they knew where the sprightly plastic ducks, frogs, turtles, and beavers were spilled. But these plastic critters may wind up telling us more about plastics degradation than how to source anonymous plastic flotsam of the kind found at a Hawaiian beach that's famous for all the wrong reasons.

Kamilo Beach is a coastal anomaly near the southern tip of Hawaii's southernmost Big Island. It boasts the attributes of a world-class tourist destination—mist-shrouded mountain backdrop, crescent bay, tide pools carved from lava, murmuring surf, and what appear to be sandy beaches. But it's a literal dump, and a place to gather evidence as well. Kamilo is one of nature's perverse wonders. Its unique location relative to north–south currents sweeping down the Big Island's leeward coast enabled Kamilo to provide more than fish for ancient Hawaiians, who well knew the specialness of the place. It blessed them with treasures from far away. Enormous logs from the Pacific Northwest were borne on westerly currents to Kamilo, there to be fashioned into canoes. Shirley Gomes, a lifelong island resident, told me that she and her friends went there in the 1940s to select wood flotsam to shape into surfboards. They also collected Japanese glass fishing floats, now rare and valued, for playing a "fortune-teller" game. Kamilo was also a place where the bereaved came to find the bodies of loved ones lost at sea. What greets you now are plastic spray nozzles, product bottles, shoe parts, Nestlé coffee lids, toothbrushes, butane lighters, fishing nets by the ton, and a few other objects that invite forensic analysis.

Kamilo is such a trash magnet because of a perfect storm of geologic, atmospheric, and oceanographic conditions. First, there's Mauna Loa, a dormant volcano so big that it splits the trade winds into prongs. Kamilo lies on the mountain's lee side and happens to be where the wind prongs reunite. Add in cross-currents and proximity to the southern border of the North Pacific Gyre, with its regular inputs of long-lived debris from Pacific Rim countries. Now you have a sub-vortex filled with trash, poised to expel it beachward.

The trip to Kamilo requires a jarring off-road hourlong trek over barely there trails pocked with dust pits and sharp lava. Most people find themselves bouncing out there for organized beach cleanups. I join a pilgrimage to Kamilo with the oceanographer Curtis Ebbesmeyer and his beachcombing buddy, Noni Sanford, who, with her husband Ron, serves as guide. She belongs to Ebbesmeyer's Beachcombers' and Oceanographers' International Association and travels from her home in Volcano Village to the network's annual meetings at mainland beach-combing hotspots. She also manages Volcano's volunteer fire and rescue service, but she's locally famous for her clever jewelry and hangings made from beach glass and plastic bobbers found at Kamilo. She's also a beach cleanup activist. Sanford corroborates other locals when she says that a few years earlier the trash was piled ten feet high in spots. She began calling county and state officials, urging action. Then she called the newspaper. Finally, she got "a threatening call" from an official who told her to back off because the message—that Hawaii's beaches are trashed—could hurt tourism. But her and others' pleas were finally heeded when the U.S. Geological Survey mounted a massive cleanup in 2003.

During the first official cleanup, Army helicopters ferried out more than fifty tons of beach trash, mostly plastic, mostly fishing gear. Now cleanups happen several times a year, organized by the Hawaii Wildlife Fund and ad hoc groups. Dozens of volunteers bring and fill many dozens of large garbage bags. But even daily cleanups would never rid the beach of trillions of tiny plastic flecks that some say now outnumber grains of sand on the 1,200-foot main beach. "I take a few bags home each time," Sanford tells me. "I soak and separate them by color. Even though they're totally disgusting, they're kind of pretty, and at least the ones I get won't be eaten by a fish or turtle." We gaze into a handful of mostly plastic specks scooped up from the tide line. Most are asymmetrical, your standard shards, but our practiced eyes instantly spot shiny little spheres among them. These are nurdles, the preproduction plastic pellets. Of all the crazy things we find at Kamilo, these make the least sense. Kamilo should be well beyond the reach of the few U.S. chemical plants that make and the many converter plants that process these pellets. Most plastic things are formed by melting and molding—be it a pocket comb, hot tub, or ketchup bottle—and nearly all begin life as resin beads. But what is "thermoplastic feedstock" doing here? Midway along the high-tide strand line, I collect 2,500 plastic fragments larger than one millimeter in a one-foot-square area. Five hundred of them are nurdles, enough raw feed stock to form a plastic bag.

Given the beach debris rule of thumb that remote beaches collect mostly fishing gear, not urban debris, nurdles at Kamilo don't fit. Is it just a fluke? After I spoke to her marine science class, Karla McDermid, a professor at the University of Hawai'i at Hilo, decided to study beach plastics. Teaming with Tracy McMullen, a senior honors student, she quantified plastic debris at nine beaches throughout the Hawaiian archipelago. They found "small-plastic debris" in all samples taken at all sites. But the greatest concentration was at the most remote beaches on Midway and Molokai, neither of which teems with commercial activity. They sieved debris from standard transects and found 72 percent of it by weight was plastic. Of 19,100 pieces recovered from 20 square meters, 11 percent were nurdles. How do you apply debris forensics to anonymous nurdles so far from home?

We know most of the 300 billion pounds of plastics produced each year start out as pellets. The pellets travel in railcars, tanker trucks, and land-sea containers to molding factories throughout the United States and the world. If a tenth of a percent of this feedstock were to escape to the oceans, that's a potential 150,000-ton annual deposit. At 25,000 pellets per pound, that's 7.5 trillion little units, about half of which would float. We're not seeing evidence of loss on this scale in gyre trawls. We look again to the work of wildlife biologist Peter Ryan, who found in the 1980s that nurdles had become de facto staples in the diets of several seabird species. He followed up in 2008 by analyzing later ingestion studies from 1999 to 2006. He found plastic ingestion rates hadn't changed, but pellet content had decreased by 44 to 79 percent in all sample groups. His conclusion: "There has been a global change in the composition of small plastic debris at sea over the last two decades." You could postulate that most nurdles on remote beaches are vintage, having tumbled into the marine environment before land-based processors cleaned up their acts and shippers developed improved methods for securing ocean-going containers.

Three exotic items commonly found at Kamilo could be featured in a "name-that-thing" quiz show on TV. The first object would be a translucent plastic cylinder about six inches long, often with a loop at one end, mangled. This one is a sop to young contestants: spent Cyalume glow sticks, patented in 1973 by American Cyanamid, popular at raves and rock concerts, indispensible to night-maneuvering military troops. In fact, the U.S. Navy controls most of the patents. They're also a basic tool for commercial fishermen, who string them by the hundreds to long-lines and nets. When you bend a new stick, a thin glass ampule inside breaks, releasing a reactive chemical and resulting in a chemical glow. Fishermen know certain fish species are drawn to light, including blue-chip catch, tuna and swordfish. In China, the motto might as well be: "Let a billion light sticks glow." Commercial fishing light sticks are a hot category, promoted on one website as "highest quality commercial fishing light sticks 4" and 6" for tuna fishing vessels and longline fishing boats over the world." A sense of the vast numbers produced can be gleaned from online sales sites. One random manufacturer among dozens—Zibo Dexing Industries Co., Ltd.—claims to make 100 million units per year and charges in the low double digits (cents) per unit. Light sticks aren't advertised as disposable, but they can't be reused or recycled. A spent light stick should be bent but whole. The ones we find are literally chewed. You have to pity the creature that gets a snoutful of these chemicals after taking a bite, along with indigestible plastic and glass. A paper in the journal *Environmental Toxicology and Pharmacology* found light stick chemicals to be "toxic to marine organisms, especially under low dilution conditions or direct contact." Light sticks are an albatross chick staple. Reusable LED rod lamps have come on the scene and are used by evolved fishermen, despite the higher cost.

The second item is a black plastic cone-shaped basket that looks like a small dog muzzle. These curious objects are used to trap a curious quarry: the hagfish, also known as "slime eel," a scaleless, jawless, charmless, mucous-coated scavenger that looks and swims like an eel and feeds on benthic carrion. South Korea—where hagfish, like oysters, are a reputed aphrodisiac— is the main market. Korea overfished its own hagfish stock, but

its market still wants nine million pounds a year. Hagfish skin
has commercial value as well, and is used in "eel skin" wallets
and various pocket items seen in urban kiosks. Korean hagfish
buyers pay U.S. West Coast fisheries up to $20 a pound to catch
them. The cone traps are poked into baited cylinders set on the
seabed. The hagfish assumes this contraption is a dead fish,
plunges its face into the cylinder, and gets snagged by spikes at
the end of the cone when it tries to get out. The traps on Kamilo
and other island beaches likely come from the old Korean fishery: most look degraded. New traps are said to be made with
biodegradable plastics. We still collect many in the gyre and at
Kamilo, relics that remind us that yesterday's plastics may linger
well into the future.

The third mystery items are black, blue, gray, and green polyethylene tubes, from one to eight inches in length and about
three quarters of an inch in diameter. These are oyster spacers or
"pipes" used in longline oyster aquaculture. Workers seed monofilament lines with infant oysters ("spat") and use spacers to
separate them so they don't clump together as they grow. The
lines are dropped from rafts anchored in protected near-shore
areas. Oyster farms are common in Asia, the United States, Europe, and other places where temperate coastal waters find shelter from surf, industrial pollution, and heavy human traffic. But
which region was launching spacers by the millions into deep
water? A likely suspect was Japan aquaculture, but their texts
recommend using bamboo spacers. China produces 82 percent of
the three-hundred-million-ton annual oyster harvest, and Korea
and the Philippines also have industries. But the literature suggests these farms would use the cheaper knot method to separate
spat. West Coast oyster farms use plastic spacers, but they tend
to be smaller-scale operations that wouldn't account for the ex-

treme ubiquity of the tubes. This was the second-most abundant item, after fishing floats, counted by Dr. Ebbesmeyer at French Frigate Shoals, and it's always a top item in the Kamilo Beach debris tallies.

A forensic breakthrough occurred in June 2010 when I was invited to be keynote speaker at an international marine conference held at the Hilo campus of the University of Hawai'i. Another speaker was a researcher from Japan, Koji Otsuka of the Japanese Institute of Technology. His report concerned access by first responders to port towns in the wake of a natural disaster. As it later turned out, his worst-case scenario model was less extreme than the historic disaster that struck nine months later, in March 2011. Nonetheless, his talk presciently described how a major earthquake would cut off access by road and air by splitting roads and runways. The other option, he said, would be access by sea, but it could be impeded by debris, including the breaking apart of oyster farms. Aha, I said to myself. This fellow may have the key to the oyster spacer mystery. During the Q & A session, I asked what types of spacers are used at these oyster farms. Bamboo? And he said sometimes bamboo, mostly plastic. I asked him if he was aware these spacers are everywhere: on remote beaches, in the gyre, in the bellies of dead albatross chicks. His English was rudimentary but I could see he was making the connection between something he'd never given a second thought to, oyster spacers, and the talk I'd given an hour or two earlier about plastic trash and its impact on the oceans. At the end of the session, after a few more presentations, the gentleman asked to speak before us all. In halting English, he vowed to take the message about plastic oyster spacers back home. He saw the problem clearly now and seemed a little overcome with emotion. I told Dr. Otsuka that going back to bamboo would help.

Unfortunately, that didn't happen soon enough. Post-tsunami assessments report heavy damage to oyster aquaculture operations centered near the cataclysm, and the release of millions more spacers.

Debris forensics can also tell us things about where marine creatures roam. Their diets can be indicators of ecosystem health or malaise. An example would be a landmark study from 2009, mordantly titled "Bringing Home the Trash: Do Colony-Based Differences in Foraging Distribution Lead to Increased Plastic Ingestion in Laysan Albatrosses?" Its five authors, including Cynthia Vanderlip, used monitors to track the foraging patterns of two colonies of Laysan albatross, one from Kure Atoll, the other on urban Oahu, some 1,500 miles southeast of Kure. The study set out to compare the contents of regurgitated boluses of fledging chicks in each group. The results would indicate which hunting grounds were most polluted, and by what. The researchers found that 100 percent of the boluses contained plastics. But the signal finding—termed "shocking" by lead investigator Lindsay Young of the University of Hawai'i—was that the boluses on remote Kure contained up to ten times more plastic than those of the Oahu-based chicks. The study found that parent albatross forage closer to home than non-breeding pairs in order to be closer to their chicks. For Kure birds, that meant heading both north to the "debris-rich" Convergence Zone and west into the heavily polluted Western Garbage Patch between Hawaii and Japan. The Oahu birds, on the other hand, headed east, to a comparatively clean area south of the Eastern Garbage patch. The irony, of course, is that the albatross from the most populous island were eating less of man's trash. The types of plastics were different as well. Most objects in the Kure boluses were marked with Asian characters. Aside from a nativity scene's worth of plastic toy figures and a sealed jar of face cream, the greatest amount of non-natural material found in the Kure boluses came, unsurprisingly, from fisheries: chemical light sticks, oyster spacers, monofilament line, disposable lighters (used to mend line and light smokes). The Oahu group regurgitated comparatively innocuous scraps and non-fishing-related debris.

There's little doubt that some Kure albatross harvested plastics from windrows, which are fake all-you-can-eat buffets on the ocean surface and as such beg forensic attention. The most massive windrow I've ever seen materialized on the 2002 Honolulu-to-California voyage, after the layover at French Frigate Shoals. But first, some background.

Technically, windrows are formed by Langmuir cells, named for Irving Langmuir (1881–1957), a Brooklyn-born General Electric lab director who won the 1932 Nobel Prize in chemistry. In his spare time he enjoyed decoding natural phenomena. On a 1938 passenger ship crossing of the Atlantic, he spotted his first windrow. It was in the Sargasso Sea—foam and sargassum seaweed arrayed in parallel rows. Captivated, he found that the phenomenon had yet to be explained and resolved to do the job himself. He devised experiments and conducted them on Lake George, near GE's Schenectady lab. What he found was that in certain conditions, the shearing force of wind blowing across calm surface water begins to generate convection cells. Convention cells occur when fluids or gases of differing temperatures collide. We see them when ice cubes are added to hot water and the water coils. On the sea surface, the result is something like logrolling, only the "logs" are long parallel tubes of water spinning at the sea surface in opposite directions and leaving flotsam. The natural seaweed and foam windrows Langmuir saw on his Atlantic crossing would now be speckled with plastic debris dredged up from surrounding waters. Windrows put the ocean's dirty little secret on display. A foraging albatross would be thrilled to come upon a windrow.

We chased the Great Windrow of 2002 and we explored it, diving around and beneath it by day and at night with underwater flood lamps. We never found its end, but we did witness the creation of mixed ghost net "boluses" of the type that so frustrate NOAA debris hunters. It was Curt Ebbesmeyer who first told me that the ocean is always knitting and weaving things together. I observed this mechanism in action as a length of rope coiled around and through a piece of trawl net. All along the windrow, ghost net boluses were in various stages of construction, from tangled little nests to massive, undulating swaths. I hadn't till then suspected windrows played a part in managing marine debris, only assumed they exposed it.

Here are items we've recovered from windrows: pipes; blue plastic tarps; plastic sheeting; laundry baskets and crates (used by commercial fishermen to sort catch and store bait); plastic foam floats, hollow plastic floats, and once in a while glass floats; footwear; jerry cans; felt pens; golf balls; glue sticks; hard hats; toothbrushes; coat hangers; TV cathode-ray tubes; tool, camera, and brief cases; fishing lures and hooks; soap, bleach, and condiment bottles; Popsicle sticks; toys; sporting equipment; umbrella handles, which I collect; scraps of plastic and balls of netting; bottle caps, of course; light sticks; oyster spacers; lighters; hagfish traps; balloons; a plastic knife scabbard; a Japanese traffic cone; plastic chair parts; and a one-of-a-kind find: a "washlet" seat—a low-budget Japanese bidet, we later learned. Many macroplastics bear Asian lettering. Fishing gear makes unfortunate sense, but Japanese and Korean consumer goods don't. These countries have efficient waste management systems. Few of these objects are swallowed whole by marine creatures, but virtually all will be nibbled and consumed as they degrade into smaller pieces.

As far as I know, polluted windrows have yet to be scientifically studied. It would be hard to do so because they appear unpredictably and rarely last more than a day or two. But their contents offer an enticing cross section of marine debris, a sense of its origins, and an opportunity to do some surface cleaning.

Bruce LaBelle, head of the California Department of Toxic Substances Control's Environmental Chemistry Laboratory in Berkeley, has quantified what we need to know, but usually don't, if we hope to trace sources of anonymous plastic debris in the oceans. We need the analytic capacity to decode the individual signatures of a piece of plastic, but plastics first need to be equipped with embedded coding, or "DNA." These markers should tell us where the base polymer resin was made, modified, and formed into a product, and what companies brokered and sold the finished product. Unfortunately, as of now, DNA typing a drop of blood is far easier than unwinding the lineage of a plastic shard. Even if we had this ability, we're still left with the difficult task of identifying entities responsible for improper disposal. Which is why single-use plastics—especially those made of thin flyaway film and foam that so readily escape human control—need to be reengineered to biodegrade in water. Or banned.

What do coastal areas tell us about the sources of plastic marine debris? The Ocean Conservancy (OC), an industry- and government-funded nonprofit based in Washington, D.C., seeks answers to this question on a single September Saturday each year when it stages its International Coastal Cleanup. The current data set comes from the twenty-fourth annual event, held in 2009. Nearly 500,000 volunteers in 108 countries and 45 U.S. states (a record number) collected and tallied an "astounding" 7.4 million pounds of trash along more than 17,000 miles of coastline. They also disentangled more than 300 birds, fish, turtles, and mammals.

The OC is sometimes knocked for producing data too sketchy to address the complexities of ocean pollution, and for projecting a "feel good" illusion that annual cleanups are a remedy of sorts. But the OC claims its goal is to furnish a "snapshot" of coastal debris as a way of promoting greater awareness, more responsible behavior, and improved policy. The top ten items—each individually counted—tend to stay similar from year to year and are mostly beachgoer trash. Cigarette butts, each one tallied, always top the list, and 2009, when volunteers collected 2, 189, 252 of them, was no exception. The next most pervasive items were plastic bags (1, 129, 774), food wrappers and containers (943, 233), caps and lids (912, 246), and plates, cups, and utensils (512, 516). The biggest jump was made by plastic beverage bottles, up three points to 9 percent. Then came stirrers and straws and, finally, paper bags. The top ten debris items were 80 percent of the total collected worldwide and added up to 8,229,337 individual items out of more than 10 million itemized and counted over the entire cleanup day. Even though only bags and bottles are specified plastic, it's safe to say the vast majority of straws, cups, plates, lids, caps, and cigarette butts are also plastic items.

This is a "snapshot" in which beachgoers are believed to be the red-handed perps. Let's compare with Kamilo, where the top category is far and away always plastic caps and lids (30 percent), compared to the OC's 9 percent. At Kamilo, you'll find way more butane lighters than butts. More oyster spacers than plastic bags. More light sticks than plastic straws. One area of commonality is plastic bottles: both have lots. But the OC will never have in its tallies what could be a million times more common than cigarette butts, and that is grains of plastic sand like those that may outnumber natural sand at Kamilo. Also, at Kamilo, virtually none of the trash is from beachgoers, who are few and far between.

And while these OC cleanups are clearly well intentioned, I sigh to myself when I read the beautifully produced reports and see that Coca-Cola, Solo Cup, Glad, and Dow are corporate sponsors, and read repetitive rhetoric that is well represented by this statement from one of the sponsors' "corporate responsibility" officers: "It will take a combined effort of corporations, individuals, and organizations *to educate people about the impacts of their behavior*" (my italics). It's this kind of moralist scolding from an uber-perpetrator—the one who passes out the weapons and then blames "irresponsible people" when someone gets hurt—that almost stirs in me a sense of wonder for its sheer audacity. In any case, I think it's a great shame that for many young people, their first and perhaps only experience of the sea and shore is as a place where they go to remove trash as part of an organized activity.

More than most states, California takes environmental stewardship seriously. By the early 1990s state officials felt that point-source polluters like factories and sewage plants had been adequately controlled. So they turned their attention to problems like nutrient runoff—nitrogen and phosphorus leaching from golf courses, gardens, and agriculture—and beach trash. Not only was the marine ecosystem put at risk, coastal towns were suffering economic loss from trashed beaches and contaminated water, and maintenance was costing too much. My favorite public agency, the Southern California Coastal Water Research Project (SCCRRP), was charged with conducting science that would guide new policy and result in cleaner shore waters and beaches.

I had a role as a consultant to a project designed to assess litter on Orange County beaches. A collection protocol was needed, so we devised a plan in which certain sections, or "transects," of each beach would be mined for debris. Volunteers were trained to collect and sift through five-gallon buckets of sand collected from these transects. The findings defied all expectations. More abundant than cigarette butts, plastic bags and wrappers, and water bottles were . . . nurdles. Based on the volunteers' counts, data crunchers extrapolated that all the beaches in Orange County could harbor more than 105 million resin pellets, weighing more than two metric tons. Which is not to say the study didn't also find normal recreational beach trash. That was no surprise. But the nurdle finding was a complete shocker. At that time we had no idea where the nurdles came from. Were they a point-source pollutant rolling out of local factories? Or were they a non-point source pollutant spat out by the ocean? Debris forensics were desperately needed, and an unlikely sleuth stepped up to the plate . . .

The year: 2002. Every pollution law in the book has been broken. The investigator turns out to be a very young scientist whom I meet by chance—due, of all things, to nagging maintenance issues with *Alguita*'s refrigeration system. The system's designer is Randy Simpkins, a former NASA engineer with a shop in Newport Beach. With his background in science, he's naturally interested in my research. On one of my visits he mentions that his middle-school-age daughter, Taylor, is in the market for a science fair project. Do I have any ideas? As it happens, I do. I suggest she pick a section of beach and monitor nurdle numbers over a period of time. It's worth knowing if deposit rates vary. A pattern might emerge, and possibly clues pointing to sources. I decide to check with Taylor eight years later to catch up with her and get her side of the story.

Taylor is a fellow marine mammal. She grew up swimming,

snorkeling, and surfing, and describes herself as "all-around in-
fatuated with the ocean." By age seven, she tells me, she knew her
life and work would be about the ocean. About a week after her
dad asks the science fair question, he brings Taylor to meet me.
In Taylor's words: "It was then that my nurdle obsession began."
We form a hypothesis and Taylor stakes out spots along the
wrack (the high-tide line) on both sides of the mouth of the Santa
Ana River. The Santa Ana is another of several concrete "rivers"
that channel rain (and debris) from the Southern California watershed to the ocean. Taylor's collection sites are one square
meter and about three centimeters deep, to be processed before
and after rainstorms over the course of a year, with Dad providing transportation. She recalls that she put each sample "in a large
bucket and transported it back to my house, where I sifted the
nurdles out with a large and small grid sifter." Taylor finds the
usual debris besides nurdles: water bottles, cigarette butts, and
plant material. But her ongoing data analysis reveals a "significant" increase in the nurdle count after rainstorms. Her conclusion: if nurdles were washed in from the sea, a rainstorm would
not trigger higher deposition rates. But if they're spillage from
careless factory handling that goes into gutters, then storm drains,
and then to the sea, a rainstorm *does* mean a nurdle spike.

Thus Taylor becomes the first researcher to definitively prove
the direct source of preproduction plastic resin pellets: the injection molding industry. And she's activated. She embarks on a
gutsy "undercover investigation" that takes her, and Randy, to
several injection molding plants in the Santa Ana River watershed. She uses the pretext of a school report about plastics manufacturing to gain entry. If they let her in, she checks "the cleanliness
of the facilities . . . or lack thereof." Many turn her down.

Taylor's project, titled "The Plastic Industry's Dirty Little

Secret," wins the local science fair and sends her to the state level. She places second in the environmental studies division and qualifies for the grueling, three-round audition process for the Discovery Channel Young Scientist Challenge. She's chosen from among hundreds, one of only forty in the country. "Though I came in being the 'blonde surfer chick from California,' I defied all the stereotypes," she recalls. She pilots her team to victory in a go-cart race, powered at her suggestion by a dive tank, and makes the front page of *The Washington Post*.

Back home, Taylor, a gifted artist, goes on to win grant funds to cover printing costs of a poster she designs for display in the forty injection molding plants that lie within the Santa Ana River watershed. In both English and Spanish, the poster explains how pellet loss-prevention is good not only for the business but the environment. Only one plant owner responds to her gift, but he invites Taylor to his plant to see for herself the difference her suggestions have made: higher productivity and less damage to his "very expensive" machinery. Now Taylor is a marine science major at the University of San Diego. She's refining and expanding her nurdle study to include toxicity issues, aiming for publication in a peer-reviewed scientific journal.

Algalita and other groups have developed best practices and worked with industry on ways to contain their feedstock. The fact is, among so-called microplastics, nurdles and degraded fragments are not the only problems. There are other categories of "primary" microplastics, meaning tiny plastics produced for commercial purposes, including industrial abrasives and cleaning compounds, plastic powders used as feedstock for PVC pipes and rotational molding, and "microbeads" added as exfoliants to facial cleansers. When these go down the drain or gutter, most will sneak through sewage treatment and debris barriers and wind up in a river or ocean. Some of these resin abrasives are actually used to clean boat hulls.

Pollution by microplastics now gets serious attention, but it took years of pushing. The first marine microplastics workshop ever, sponsored by NOAA, was held in September 2008 at the University of Washington Tacoma. Among the attendees were Tony Andrady and Algalita's Marcus Eriksen. A major outcome was the consensus that serious "knowledge gaps" remain and warrant study. In 2011, the Fifth International Marine Debris Conference in Honolulu finally focused on small plastics instead of derelict fishing gear—in no small part due to Algalita's efforts to shine light on the issue.

Now let's end where we started, with the mystery of the blue spirals found by an eco-beachcomber friend on Seal Beach. My plastics industry insider friend had no idea what the little hair-curler-looking things were and why they would have wound up on Seal Beach. But he's curious too. So in his rounds as a sales rep, he makes inquiries. And he learns they're called "pigs." *Pig*, in this case, is the name for a sharp-edged little object or "scrubber" fed into pipes to scrape off occluding crud. My friend gets a tip that these pigs are being used at a Southern California Edison power plant perched near the San Gabriel River, just upstream from Seal Beach. The plant runs seawater though its cooling system, and the seawater contains algae spores that manage to cling to the insides of the pipes, and grow, and flourish, and clog the pipes. These spiral pigs are fed into the pipes to scrape off the algae, then blithely flushed into the river along with the resulting algal slurry. The flushing is done periodically, explaining the spirals' strange habit of magically materializing at what seem random intervals.

How could this be allowed? Once we've tracked the source of

the Seal Beach pigs, the Long Beach/North Orange County Surf-rider chapter hatches a plan of action. At this time, in the 1990s, federal and state antipollution laws have ambiguities. It becomes clear our best course is to invoke an international treaty, namely, MARPOL Annex V. The U.S. Coast Guard is charged with enforcing this law in U.S. waters. I have a few buddies at the Port of Los Angeles Coast Guard station and pay them a visit. I have in hand a letter on Surfrider Foundation letterhead stating that Southern California Edison appears to be discharging plastic "pigs" into navigable waters of the United States. This puts the Coast Guard on the hook to investigate and enforce. The commander begins an investigation and duly confirms that the stretch of river between the power plant and the ocean is navigable. Then he writes a letter to SoCal Edison advising them that plastic debris recovered on Seal Beach has been traced to their plant. And that the power company is obliged by law to take measures to prevent said discharge. SoCal Edison agrees to procedural changes that should prevent future pig releases. But a few still manage to wriggle free now and again and make their way to Seal Beach, where they can sometimes be found to this day, basking in the sun's degrading rays, waiting to be inhaled by the ocean.

CHAPTER 15

ERASING OUR PLASTIC FOOTPRINT

Hard truths are countered with convenient but unlikely hopes.

—BENJAMIN ROSS AND STEVE AMTER, *The Polluters:*
The Making of Our Chemically Altered Environment

IT SEEMS LIKE A RITUAL. First the man-made environmental crisis is dissected in painful detail, then comes the well-considered list of "solutions," the implementation of which will make the world right again. But the good ideas too rarely seem to bear fruit. Why? Because change is hard and powerful people and organizations benefit from the status quo. Plastics are a high-stakes game, and those who run it can ill afford to lose control of the playing field. But ridding the oceans of plastic means stopping all plastic inputs—now. Then a waiting game. Our residual plastic footprint—what can't be taken from rivers and shores or plucked from the oceans—would be left to decompose or strand on its own still-unknown timetable. Plastics exploded into our lives around the same time atomic energy was unleashed, and there's no putting these long-lived genies back in their bottles. But that shouldn't stop us from making sure the way we use them won't harm us or the natural systems that support all life on earth.

It bears repeating: waste generation in the United States nearly doubled between 1960 and 2007, from 2.68 to 4.63 pounds per person per day. We've seen how Americans during these decades became "consumers" faced with a proliferating array of product choices, and how women entered the workforce in unprecedented numbers and learned to value convenience; and we've seen how disposability became equated with convenience and hygiene and how it led to worldwide economic growth. We've seen how even costly consumer goods—especially electronics—have inched toward disposable status. Daily, weekly, monthly, and annually, as our stuff breaks, wears out, becomes obsolete, goes out of style, or, worse, becomes boring, we get the new version or the next "big thing." Few among us are exempt, and the recent reality TV stars, the compulsive hoarders, illuminate an outcome of this behavior in an extreme way. As they slowly self-destruct by releasing nothing to the waste stream, these odd souls ascribe sentimental value or potential usefulness to most anything—old or new, whole or broken, cheap or expensive—that comes into their lives. While appearing to be consumers run amok, hoarders represent the yearning for a link to eternal renewal. In truth, everything *will* be useful again. I deeply sympathize with their tenacious search for meaning and lasting value in products and tend to see them as victims of an irrational system with no sensible endgame for our precious resources.

Because our consumption patterns now result in such a huge turnover of goods, the idea of recycling has come to represent the best solution. The chasing-arrows symbol was created in 1970 by Gary Anderson when he was twenty-three (so was I) for a contest devised by the organizers of the first Earth Day. Thus branded, recycling became part of the national platform for managing municipal waste. The chasing arrows appear on most plastic food and personal product containers, but they're conspicuously absent from plastic films, foam shipping material, and those infernal, hard-to-open clear plastic packages—the so-called blister packs—that contain small pricey electronics, tools, gadgets, pop-out pills, and beauty products. The numbers 1 to 6 inside the chasing arrows indicate the base polymer type, but not all are collected in recycling programs. Number 7 simply means, "not 1 to 6." But even the "in-demand" resins—typically 1, polyethylene terephthalate (PET); 2, high-density polyethylene (HDPE); and, to a lesser extent, 4 and 5, low-density polyethylene (LDPE) and polypropylene (PP)—aren't always recyclable or recycled once they're collected. Of the number 1s and number 2s that are sorted, baled, and sold, the winning bidder is usually the one selling to concerns located where labor costs are low— typically China. If no one bids, the pallets of sorted plastics are headed for the landfill anyway. Much of recycling is an elaborate charade.

Recycling rates have not kept pace with increased plastic use. This is unfortunate in light of the cadres of entrepreneurs I encounter with all manner of ideas for converting waste plastics into profits. Their work is not easy. Plastics have proved resistant to post-use reprocessing because they can't be lumped as a single material. RecycleWorks, a program of San Mateo County in Northern California, claims there are fifty thousand types of plastic. Though it doesn't cite a source, the statistic seems defensible. Consider both thermoset and thermoplastic types, and within those broad categories, the millions of applications. Many require specific characteristics imparted by special chemical combinations: the elasticity of chewing gum, the heat and oxidation resistance of lawn furniture, the stretchiness of pallet wrap, and the strength required of carbon-fiber-reinforced polymers used to make elite racecar bodies, sails for world-class racing yachts, bulletproof vests, and the new generation of rocket ships. Separating these variations into unique streams of recyclable material is impractical. Yet new variations emerge with no end in sight. New patents for polymer-related materials average more than fifteen a week, according to U.S. Patent Office "gazettes." Many involve advanced nano materials, gels, coatings, and the like, but significant others concern laminates for packaging and new foaming techniques, the next wave in "lightweighting" packing supplies so they require less raw material, weigh less for shipping, and can be promoted as green—or at least greener.

We need complete, accurate, and, if I may say, helpful labeling to guide the sorting of complex products. But the U.S. agency in charge, the Federal Trade Commission, has no authority to move labeling in any particular direction for any social goal. It is empowered only to regulate untruthful labeling. Manufacturers can put anything they want on packages, as long as it is the truth. That means psychologically satisfying, easily defended hype words like *new* and *improved*. But there are some rules. The Fair Packaging and Labeling Act of 1966 requires: (1) the identity of the product; (2) the name and place of the manufacturer, packer, or distributor; and (3) the net quantity of the contents in metric and/or inch or pound units. It hardly seems that restraint of trade or information overload would result from including recycling or disposal options for the container. Suppose the U.S. government decided that every plastic package and product should come with clear instructions regarding its "end of life" or, better, its next stage of life. How would such a mandate be promulgated? Would it take something like the Clean Water Act, with its "Declaration of Goals and Policy," that forty years later is still struggling to be implemented piecemeal even by the most concerned states and municipalities?

In the early 1990s, Germany decided to tackle the recycling problem by stepping up separation at the consumer level. Bottles with a deposit are redeemed at the store. Non-deposit glass is sorted by color (clear, brown, green) and put in public bins provided in every neighborhood—but not late at night or early in the morning to avoid noise. Green and blue bins are furnished to homes for paper and cardboard. Brown bins are for "biologicals"— "biodegradables" in the United States. Yellow bins or bags are for packaging items sporting the green dot logo and include plastic but also aluminum and tin cans. Gray bins are for other things like cigarette butts and disposable diapers and old fry pans. The contents of the gray bins are burned after removal of metals. Germans are charged by weight for what they put in the brown, blue/green, and gray bins only. The green dot program for packaging, destined for yellow bags and bins, is paid for by the industries that produce it. They employ the trucks and drivers that pick it up and the workers who sort it. When it comes to packaging, the key concept is that the producer pays. As a rule of thumb, the recycling cost is about one euro ($1.50) per kilogram (2.2 pounds) of packaging, and is pegged to the actual cost of recycling these materials.

The green dot compliance program is now trademark registered and can be licensed for use in 170 countries around the world. There's no charge to use the green dot if the licensee contributes a prescribed amount to the recycling operations that deal with their products. In Germany, the cost to recycle these materials over a ten-year period came down 75 percent and put the country on track to make its 2020 deadline for closing all its landfills, since packaging is a major landfill component. The color-coded bin system also covers public areas and workplaces and makes it easy to do the right thing. An important aspect of recycling is uniformity. Uniformity in production. Uniformity in collection. And uniformity in mandated recycling. This uniformity is economical, yet it goes against the promotion of individuality and local control in the United States. Without large volumes of identical plastics to recycle, creating useful industrial feedstocks is difficult. It remains unclear how remelting changes desirable qualities in plastics, or if there is a remelt limit. Not to mention how much virgin plastic has to be added to the recyclate for quality control. Uses for unseparated materials are limited, so demand is lacking and so are profits.

In Germany, trash burning and other forms of thermal processing are last resorts for unrecyclables. Not so in other countries, where the practice is given names like *gasification*, *pyrolysis*, *plasma arc*, *waste-to-energy*, and *thermal recycling*. In California, carpet makers are pushing to have their chemically laden product designated a "fuel" so that burning waste carpet can be considered part of the industry's mandatory "recycling plan." In Germany, the plastics industry pushed hard for "thermal recycling" for energy generation. The issue became politically charged, but the majority of Germans stood firm behind science showing that burning mixed plastics adversely affects air quality. Not only is CO_2 released, contributing to global warming, but so are atmospheric dioxins and furans. These are among the most toxic chemicals known, even in vanishingly small amounts. A state of

Massachusetts report on municipal solid-waste handling options found that all these systems produce "air emissions containing particulate matter, volatile organic compounds, heavy metals, dioxins, sulfur dioxide, hydrochloric acid, mercury, and furans." Ecologist-author Sandra Steingraber writes in *Living Downstream: An Ecologist's Personal Investigation of Cancer and the Environment*, her book about man-made cancer-causing agents: "Even the newest, fanciest incinerators send traces of dioxins and furans into the air. . . . [A]ll of us have a stake in the question of whether or not generating electricity by lighting garbage on fire is a genuine form of renewable energy." Dioxins form when plastic is burned in the presence of a chlorine donor. The donor molecules can be organic or inorganic. For those who imagine burning plastic marine debris as a partial solution—as is happening in Honolulu and several other locations in the Nets-to-Energy program—remember this: the debris will be coated with seawater. One of the ways we get table salt—sodium *chloride*—is by evaporating seawater. Ergo, burning plastic coated with sodium chloride will produce dioxins.

Most large ships are equipped with incinerators that burn trash generated onboard, including plastics. Their smokestacks often lack scrubbers to remove harmful emissions. A policeman at the Port of Hamburg, Germany, who inspects such vessels, told me about operational problems with these incinerators. Protocols call for preheating the trash oven to ensure complete incineration to ash, legal to dump in the ocean. But operators often load the oven with trash, *then* turn on the flame. This results in lower temperatures and incomplete incineration, especially at the center of the heap. The resulting slag would not be legal to dump. I realized this practice probably accounts for the formless mixed plastic blobs that show up in my samples. The world's oceans are the preferred transport medium for the goods of rapidly growing global commerce, and they get a big double dip of air pollution from engine exhaust and incinerator smoke. Studies show increased cancer and lung disease rates among people who live near industrial ports where ships burn heavy fuels to keep their generators going while loading and unloading cargo.

There is, however, another advanced technology called chemical recycling that Germany allows for plastic waste. This system works with mixed plastics and resembles petroleum refinery cracking, subjecting polymers to high temperatures and pressure in the absence of oxygen, then adding hydrogen. The results are products similar to those derived from crude oil, including basic plastic feedstocks. Advanced systems being developed in the United Kingdom will crack mixed polymers back to their monomers, extracting, for example, styrene from polystyrene, and terephthalic acid from PET plastic, which can then be reused. The basic plan is for these plants to make fuel: low sulfur gasoline and diesel. Their emissions are contained and they're fueled by their own product. They're not perfect by "closed-loop" standards, because the energy used to make the plastics in the first place is not recovered, and there is a toxic sludge left over that must be landfilled. The United States lags behind Europe in this technology, here called pyrolysis. A number of start-ups seek investors, but none are yet operating on a large scale. Polyflow in Akron, Ohio, is one such company. I met its two young principals when I spoke in spring 2010 at the Cleveland Museum of Natural History. They had approached me in advance, wanting to know more about the Great Pacific Garbage Patch, and I realized they hoped their technology might be a solution to the problem. I get this a lot, and I appreciate the concern. While there's no question that new technologies are needed to help

staunch the flow of ocean-bound plastic trash, and there will be
a need for dealing with the mess of mixed debris from cleanups,
most of the ideas I've heard fail to address the core issue: the
ever-growing volume of plastic products and packaging deployed across the world. It's a simple fact: a percentage of these
plastic items escape, and even a small percentage means a lot of
marine debris. Industry and government-backed groups refrain
from applying pressure to the supply side of the equation because it's un-American, or anti-business. The messages Americans hear loud and clear are "Don't litter" and "Recycle," which
burden the consumer with trash management. Less loud and
clear are the two other Rs: "Reduce" and "Reuse," which carry
a faint aroma of subversion. That the bulk of our best plastic
recyclates are shipped to China also tends to be kept under
wraps. About four billion pounds of plastic a year leave U.S.
shores for recycling abroad—more than 10 million pounds a day.

The export of our plastic recyclables is not seen as a good
thing by those who would banish the term *waste*. To the resource recovery movement, every commodity has value no
matter at what stage of its life cycle it becomes unwanted. If
everything is a resource, then all we have to do is figure out how
to reuse it at the appropriate stages in its life cycle to achieve zero
waste. Thrift stores are great for stuff that still has life in it, and
they're being integrated into waste transfer stations. Though it's
easy enough conceptually to view plastic packaging and used
plastic products as resources, in practice it is often necessary for
government, industry, or both to subsidize the recycling effort.
There's just not enough profit in recovering, sorting, cleaning,
processing, and remanufacturing infinitely variable plastics. This
is why we need extended producer responsibility, so that indus-

try won't make things that it can't economically recover. Will there be exceptions to this rule? Certainly. Should there be a lot of them? Not if we're serious about erasing our plastic footprint.

All this begs the question: why is it so hard to make a new plastic thing out of an old one? We do it all the time with glass, paper, aluminum, and steel. One of the reprocessing limitations of plastic is its low melting point. In the crucibles of industry, glass, aluminum, and steel must reach temperatures so high during the melting process—in the thousands of degrees—that whatever contaminated the item—be it food, paint, or oil—is vaporized. Paper is chemically and mechanically re-pulped. Most thermoplastics, on the other hand, begin to soften and melt at or even below 212 degrees Fahrenheit, the boiling point of water. Before they are remanufactured, plastics must be carefully washed, an extra step. Even so, since we know they are lipophilic (oil absorbing), they cannot be cleaned thoroughly enough and thus are not suitable for applications that involve food contact. Some recycled plastic pots in the nursery industry still carry plant pathogens. There are thick plastic refillables in Europe, but evidence of increased leaching with age and use makes this tack questionable.

With all the problems associated with plastics recycling, it is no wonder that compostable alternatives are being developed. It should be understood, however, that using plant material to make "bioplastics" does not guarantee its biodegradability, compostability, or marine degradability. Those carbon-to-carbon bonds that form the backbone of plastics can be just as persistent in a polymer made from plants as one made from petroleum. It's all in the way you make it. Today the majority of the world's chemists are polymer chemists. Their sophistication is such that they're now able to fabricate custom polymers molecule by molecule. Many of them work in the pharmaceutical industry, creating bio-mimetic polymers that can deliver drugs to specific organs, tailored to targeted receptors. Other polymers are engineered to biodegrade at just the right time after they have done their work: think surgical thread that dissolves after the incision heals. Some companies are marketing accelerant additives to conventional polymers that make their carbon chains break apart, but not necessarily to biodegrade completely or in a timely manner. Such additives are added to six-pack rings that were shown to girdle wildlife in early shock photos. They are the chemical equivalents of the scissors people use to cut the rings before disposing of them. The fact that a synthetic polymer breaks apart is no guarantee that it will biodegrade into the molecules that originally formed it, principally carbon dioxide and water. For this reason, it is likely that many of the broken six-pack rings still remain afloat in the ocean as fragments of their former selves.

In order for a plastic to disappear in the ocean, it must be marine degradable, essentially undergoing the same processes as organic materials in a terrestrial compost pile. Just because a plastic will completely biodegrade in a compost pile doesn't mean it will biodegrade in the ocean. The ocean is much colder than a compost pile, and some biodegradable plastics require considerable heat (around 140 degrees Fahrenheit) to break down into their constituent elements. On land, the heat is generated by an exploding diversity of life in the pile: fungi, bacteria, and insects wildly proliferating. In the ocean, this process is barely operative. I find only one insect in our trawls in the ocean, a water strider, *Halobaites*. Fungi are also scarce, and found mostly in coastal environments. While bacteria and viruses are legion, they are working in a relatively cool environment, which means relatively slowly. A rule of thumb is that bacterial activity doubles for every increase of 10 degrees Fahrenheit. So it is easy to see the tremendous increase in activity going from 60-degree seawater to a 140-degree compost pile.

The University of Idaho invited me to speak on a program that included Brad Rodgers, the inventor of the first compostable consumer product package. He explained to me how his team at Frito-Lay created the Sunchips bag out of plastic derived from lactic acid fermented by microorganisms fed a corn-based diet. The bag is compostable plastic on the outside and moisture-blocking aluminum on the inside, to preserve crispiness. Another layer of plastic coats the aluminum next to the chips to prevent oxidation. Intrigued by all this, I visited the grocery store near my hotel and bought a few bags of Sunchips. They look like the rest of the chip bags, but the first thing you notice as you pluck one off the shelf is the crackle. You simply cannot touch a Sunchips bag without making a lot of noise, something like the worst static you can imagine. The chips themselves taste like a slightly sweet, less salty Frito, with the corn taste muted by other grains: wheat, rice, and oats. According to Rodgers, the chips convey a message of health and wellness, and the new bag represents "green." The combination of the two has more impact on consumers than either alone. The polylactic acid needs a "thermophyllic," i.e., microorganisms in a hot compost pile to break down. The package has degradable inks and the aluminum will mix back into the soil whence it came, being the most common metal in the earth's crust. Rodgers disclosed that if the temperature in the compost pile is not at least 100 degrees Fahrenheit, the bag is rotting, not composting. The ocean never gets over 100 degrees except near deep thermal vents. Perhaps in the not-too-distant future a Sunchips package will end up near one and there be "marine biodegraded." Creating plastic packaging that is both compostable and readily marine degradable would be another good step for Frito-Lay (and other snack and beverage companies). Marine degradable, *not* marine disposable! Rodgers says they're working on it. However, several months after the grand debut of the new bags, I read that compostable bags for all but the "original" flavor had been withdrawn from the market. Consumer complaints about the noise trumped green except in Canada, where Frito-Lay offers free earplugs on its website. Rodgers has recently reformulated the bag to eliminate the crackle, showing that consumer complaints in some cases yield rapid results, but the bags are still not marine degradable.

Metabolix is a company that has taken up the challenge to produce marine-degradable feedstocks for the plastics industry. Their process lets specialized microbes synthesize feedstocks from readily available resources like sugar, vegetable oil, and starch. The microbes are grown into a large, healthy population and then fooled into storing energy by stressing them—depriving them of normally available elements like nitrogen, phosphorous, and oxygen. They react by storing energy as natural polymers called polyhydroxyalkanoates (PHAs). What makes PHAs so promising is that they shed water like petroleum-based plastics but are readily consumed by bacteria found in the marine environment. The goal is not to create a plastic so functional and eco-friendly that it overcomes the need for responsible handling at the end of its first use, but rather to replace conventional plastics that occupy established industrial niches in agriculture, aquaculture, and fishing. For example, lobster traps. These devices require a door that will open after a certain time, freeing the lobsters if the trap is lost. A marine-degradable plastic is a good candidate for these doors or door latches. PHA is also a good candidate to replace conventional plastics used as agricultural film. Strawberry growers, for instance, use the film to keep down weeds, hold moisture, and warm the soil around their plants. At the end of the season it has to be removed. It would be much simpler if it could be tilled into the soil and broken down by soil microbes, much like paper.

There are those who see the end of cheap resources as a powerful incentive to adopt a closed-loop economy. William McDonough and Michael Braungart are an American architect and a German chemist who coauthored *Cradle to Cradle: Remaking the Way We Make Things* and advocate a design revolution that would bring about closed-loop production and consumption. The concept is that the birth of a product culminates in its rebirth as a new product. The endgame is a new game with the same deck. This cradle-to-cradle paradigm connects with an exciting set of concepts known as the Hanover Principles. These call for the elimination of the concept of waste, but also include two key points that encompass plastics pollution: (1) insist on rights of humanity and nature to coexist, and (2) create safe objects of long-term value. *Long-term value* means not only durability, but recyclability. This takes the onus off consumption as the problem and puts it instead on industrial design, which must devise recyclable components for each product. To achieve zero waste and a cradle-to-cradle society, it will be necessary for people to respect their material goods and assure their life through many incarnations by husbanding them within the new production apparatus. The problem is our products come with so much that isn't part of the product. In some cases, the product is actually worth less than the package. How can the respect one feels for something that is truly useful be transferred to the "packa-rage"-inspiring blister pack, which requires heavy shears to avoid injury to fingers prying and pulling the tough, sharp edges? Another hurdle is creating an infrastructure to reabsorb used materials into the production apparatus. This cannot be economically accomplished until we insist everything produced be expressly designed to be recycled. The final hurdle is overcoming the wasteful habits we associate with prosperity and the uniquely American delusion that directing technology toward the common good is tyrannical.

The new field of green chemistry may at last show us the way toward a less toxic future. In forward-looking universities across the nation, chemistry majors are learning a new ethos that prioritizes development of kinder chemicals to take the place of toxic ones. Green chemists are simply good chemists that start with safe chemicals and end with safe chemical products. If hazardous substances are absolutely necessary to create an ultimately safe product, they are handled on a "lease basis" and returned at the end of the manufacturing process. They will not be delivered to the consumer.

Let's look more closely at how plastic products are manufactured. As noted earlier, nurdles are produced by catalyzing gases and liquids in chemical plants. The second step is to send the little pellets to a "converter" or production facility that will convert them into a product. Depending on the type of product, the factory will be equipped with machines that melt the pellets, then blow-mold them, roto-mold them, extrude them, spin them into fibers, or flatten them into film. But the final product will need specific characteristics, attributes like color, transparency or opacity, sheen, flexibility, hardness, UV resistance, and heat tolerance. So the basic polymer feedstock will be mixed prior to or during melting with a suite of appropriate additives, mostly chemicals known as monomers, stand-alone molecules. Thousands of chemical additives are in current use and labs around the world modify and develop hundreds more each year. As we've seen, many are toxicants and endocrine disruptors, which can leach out into whatever they contact. We should not be forced to choose between utility and safety, but since there are few standards for these compounds, it is up to the converter to choose what works most cost-effectively for their application. A small operator would have no way to determine the safety or risks of the additive except by the claims of the vendor. Green chemists are developing a new generation of additives and plasticizers that impart the desired qualities without rendering the material toxic or difficult to recycle.

Capturing wayward plastics before they reach the ocean is a basic way to reduce marine pollution. Since the ocean is downhill from nearly everywhere, drainage systems for cities and towns typically lead to rivers that, however circuitously, eventually empty into the ocean. And let's not forget the windblown component. Because of a high strength-to-weight ratio, one might even say utility-to-weight ratio, very lightweight plastics have many disposable uses, especially in food delivery, and can easily blow into the ocean. Just as municipalities have evolved to be the responsible entities for dealing with our sewage, so, too, they are charged with managing our solid waste. The federal Clean Water Act of 1972 is a "Congressional Declaration of Goals and Policy," which sought "restoration and maintenance of chemical, physical and biological integrity of the Nation's waters" and had the optimistic goal that "discharge of pollutants into the naviga-

ble waters be eliminated by 1985." First, I think we must ac-
knowledge that plastic in the ocean, lakes, and rivers is a *pollutant*
that compromises their *physical integrity*.

The paved "hardscape" of a major U.S. city sits atop a laby-
rinth of storm-water runoff conveyances. Such systems are rare
in much of the world where the consumer plastic lifestyle has
been embraced. Even in places where infrastructure exists, the
cost of catching debris can be prohibitive. Runoff in less developed countries follows the contours of the landscape to streams
and rivers or blows across deserts. Photos abound of individuals
in small vessels navigating through floating fields of trash in
Asian rivers, harvesting the marketable types of plastic waste. If
it were sewage that was the pollutant in question, no one would
suggest we disinfect it after it has been discharged from the end
of the pipe into the ocean. But so-called end-of-the-pipe solutions are commonly prescribed for plastics pollution, probably
because it is massive, solid, and visible. Beach cleanups organized
and executed by volunteers have become the most common
group activity on beaches worldwide, surpassing surf and beach
volleyball contests. What was a once a yearly activity promoted
as the Ocean Conservancy's Coastal Cleanup Day has become a
year-round activity, happening monthly in some places. The
cleanups are international in scope and have expanded to harbors, lakes, rivers, and upstream areas. Now that five major high-
atmospheric-pressure gyres have been identified as accumulation
zones for plastics that escape these exercises, groups are forming
to go out and clean them up.

Some recycling entrepreneurs envision a system that would
vacuum up plastic trash from the oceans and reprocess it into wonderful products and good press. Some have developed fanciful

devices—even trash-powered "islands." One group in Rotterdam, the Netherlands, drew an amazing tentacle island with nets pulled along the tentacles by parasails, bringing plastic trash to a central processing unit resembling the body of a starfish. There are two main reasons why these types of solutions are exercises in futility. The first is the enormous size of the floating plastic convergence zones, and the second is the dispersed nature of the debris. Forty percent of the world's oceans lie within subtropical gyres. These are convergence zones totaling 145 million square kilometers. All the land mass on earth is only five million square kilometers larger than the gyres. A trash-to-energy plant typically burns one ton of waste to produce 550 kilowatt hours—enough energy to run a typical office building for one day. We have evidence that the gyre's plastic content is double the amount we found in 1999. That means ten kilos of skimmed debris per square kilometer, one of which might take several days to trawl. You'd need to trawl one hundred square kilometers over several months to harvest a single ton of fragmented plastic material to power an office building for a day. There are large ghost nets and floats out there, too, perhaps an average of one large item per square kilometer in areas of high concentration. These mega-plastics could up the take considerably. In a week's time onboard *Alguita*, we are able to dip net or haul out about five hundred kilos of plastic. In this scenario, skimming and netting for a week could produce a day's worth of fuel for our office building. It still doesn't pencil out.

So let's look at cleaning up the gyres with no commercial motive. We now know all are similarly polluted with plastic trash at concentrations like the Eastern Garbage Patch near their ill-defined, shifting centers, and that the plastics thin out toward the edges. But since amounts are increasing rapidly, let's assume, for the sake of argument, that they are homogeneous at ten kilos per square kilometer. Remember, land-based plastics are on ocean current "highways" to these gyres, and the highways run outside the convergence zones. Nevertheless, for the cleanup we start with the 145 million square kilometers that comprise the subtropical gyres. And, to be generous, let's say an advanced cleanup vessel can do 5 square kilometers per day (10-meter-wide net traveling at 20 kilometers an hour, wider and faster than anything in use today). It will only take that vessel 29 million days, or 79 thousand years to do the job, or, for grant-writing purposes, a thousand boats 79 years working 24-7. This is surface debris only, and we don't take into account the associated organisms that would be destroyed. A boat with this capability would cost thousands of dollars a day to operate. Moreover, new research presented at the recent Fifth International Marine Debris Conference in Honolulu suggests that 90 percent of plastic debris was located below the surface layer, mixed into the water column, when the wind blows at only 10 knots. Does this mean we should do nothing, that the task is impossible? No and yes. These ocean-borne plastics are killers, and when they hit sensitive habitats, like the Northwestern Hawaiian Islands, they destroy rare coral forests of unimaginable diversity and beauty. As we've seen, an average of 52 tons of derelict fishing gear and ropes impact this area every year and have to be carefully untangled, cut off, and removed at a cost of millions of taxpayer dollars. Because of this, the U.S. National Oceanic and Atmospheric Administration studied the possibility of capturing these huge ghost nets at sea before they impact the fringing reefs of the coral atolls under their jurisdiction. Chapter 12 described our tracking buoy deployments in support of this effort. But a deeper look at this effort serves to illustrate how even our most high-tech tools are no match for drifting rafts of plastic. NOAA scientists identified two ocean features that are easy to measure from satellites, and correlate well with debris concentrations, the sea surface temperature, and the blue-green chlorophyll produced by phytoplankton. They then flew over the areas they predicted would have debris concentrations and, sure enough, there they were. They used this confirming information to create what they called the DELI Index. DELI stood for debris-estimated-likelihood index. The area where they applied DELI was about a million square nautical miles (3.4 million square kilometers) and showed puffy cloud-shaped accumulation zones with smaller hot spots. The next step was to go there and get the stuff with the help of drone aircraft. Their model was predictive for late spring. Between March 24 and April 9, 2008, the NOAA ship *Oscar Elton Sette* (twenty to thirty thousand dollars a day) cruised a thousand miles north of Honolulu to the area where they expected to find and retrieve the ghost nets. They had at their disposal *Malolo*, an unmanned aircraft built by Airborne Technologies with a seven-foot wingspan, cameras, and sensors, capable of flying a preprogrammed pattern or being controlled by an operator on the mothership. They launched the drone by hand-throwing it off the bow and retrieved it from its ocean landing spot in an inflatable dinghy. The drone was programmed to sense color differences on the sea surface and thereby detect and note the position of large masses of debris, which would then be retrieved. They had bad luck. It was foggy in the target area and they couldn't see anything from the aircraft. The giant binoculars on the flying bridge of the ship were also nonproductive. One large hawser, not a net but a rope, was their only significant debris find. The question they wanted to answer was: what does it take to remove marine debris at sea? The answer came back: more than we've got.

Few would disagree that the proliferation of cheap plastic stuff—a by-product of the global economy—is at the core of the ocean's problems with plastic pollution. Because large organizations must embrace that economy in order to obtain the funding necessary for their operations, it is futile to look to their programs for disengagement with the plastic-plagued economic paradigm. Groups like the Ocean Conservancy and Project Kaisei solicit partnerships with companies like Dow and Coca-Cola to fund beach and ocean cleanups. I'd be somewhat appeased if they would at least pressure Detroit (which I use as a metaphor for wherever cars are made nowadays) to install trash compacters in the eco-friendlier new models. Yes, mini trash compactors. Since automakers nixed ashtrays, there is no place to put all that take-out trash except in a bag on the floor, which quickly fills up and too often is chucked out when annoyance overcomes guilt.

Forward-thinking individuals and small groups have been the first to daringly disengage from the lifestyle driven by plastic junk. Artists are often at the forefront, because they seize on waste plastic as an easily available material to play with. Consciously or unconsciously, their works abstract from and thereby break with the status quo's oppressive pseudoprosperity based on gadgetry. In this way they are prying open a space where the re-visioning of plastic can commence. Judith Selby and Richard Lang picked up plastic trash on Kehoe Beach near Point Reyes, California, on their first date and began making art and a life together out of plastic flotsam. They decorated their car with it, got married in clothes made from it, and have a barn full of it, collected and sorted by color during the first decade of the new century. They do installations, such as the one at Stanford University in 2009 called *Disposable Truths*. "We have all been sold the myth of disposable plastic. We throw it away but it stays with us for centuries and may ultimately irreparably altar [sic] the planet." They wish to "excite the aesthetic sensibility . . . in order to demonstrate the ubiquity of plastic waste in our oceans."

Pam Longobardi is a professor of art at Georgia State University. She has photographed and collected plastic trash from remote sites in Hawaii and done installations of her work in Italy, China, Slovakia, Poland, and other countries. She described two stages of feelings after visiting Kamilo Beach and viewing its heaps of plastic debris: "First, there is a spark of pure retinal pleasure: multitudinous bright colors and bold shapes of plastic objects, pile upon pile of tangled textural knots of drift net, in scales that boggle the mind, from golf cart- to whale-sized. A split-second later the mind processes what the eyes see and, with a sickening thud, there comes the recognition that it is our garbage, the residue of our everyday life, the quotidian castoff."

Andrew Blackwell is an artist who picks up debris on the beach near Brisbane, Australia. He has made sculptures of surfboards using broken bits of plastic and also abstract sculptures using broken surfboard parts he found washed up on the beach. His "Banal Enviroscape" is "but a snapshot of consumer culture's plastics alienated from its culture and often left to navigate the oceans until landfall at a beach near you or even trawled from the top surface layers of waterways." I used photographs of his surfboards in presentations to Surfrider chapters and the Institute for Figuring, which makes community crocheted coral reef installations. These artists have all contacted me to let me know that my research was an inspiration in their work.

The news media keeps pace with the stories that they think will inspire and interest viewers, readers, or listeners. Newspapers came first as we publicized our research voyages and what we found, then documentary filmmakers, then radio, then magazines, then television.

Once artists have created the abstraction that reveals the awful reality, and the media has done its "reveal" with the help of scientists and experts, it is up to activists to initiate change. What *needs* to be done battles with what *can* be done within the political given, and movements to contain the plastic monster, small but widespread, develop. Individuals like Beth Terry, who started a personal quest to live plastic-free and encourages others to do the same on her website My Plastic-free Life, feels that she has found her "voice" in the campaign. She weighs plastic discards to get the amount of disposables that have come into her life each week. Her goal is to buy no new plastic items including anything in plastic packaging. What product challenge will she work on first? Getting milk in bottles, bread in paper, local vegetables without packaging? In spite of the admitted inconvenience, she has recruited more than one hundred people from around the world to take her challenge, listing each week's plastic, weighing it if possible, photographing it, then filling out a questionnaire. The community exchanges ideas and devises tactics for ridding their lives of plastic.

The tipping point may still be distant, but a consumer groundswell against plastics is the most potent weapon in the change agent's arsenal. In my decade and a half in the trenches, I've seen tremendous growth in awareness. But awareness must become resistance—even rebellion—if the status quo is to be changed. In the United States, where too many people with political clout view environmental restrictions on business as akin to burning the American flag, the movement is necessarily diffuse and often focused on local policy. As we've seen, large product manufacturers will paint themselves green by tweaking their packaging and production systems. Just don't ask them to take back their trash or pay for its disposal. But they've learned there's green, the monetary kind, in touting the greening of their products and packaging. Smaller regional producers are making headway by responding to the demands of environmental activists with reusables and local solutions that eliminate throwaways. One by one, communities are banning thin shopping bags and foamed polystyrene fast-food containers. Will fragmented regional efforts coalesce? Not anytime soon. California's recycling rate for redeemable bottles and cans is 82 percent. Mississippi has no redemption program and recycles only 13 percent of its bottles and cans.

I believe the desire in Beth Terry and her conspirators to refuse plastic junk is driven by an instinctual aesthetic and non-ideological morality that rejects the ugliness and stress of the hyper consumer society and its mindless waste. In natural settings, everything is used and recycled seamlessly. It's a beautiful system, and it may be at the heart of our aesthetic instincts. What is beautiful is not arbitrary, nor is our feeling of solidarity with the rest of humanity, and the biosphere. The economies of modern societies, driven by the need for growth through increased sales, are not at all seamless in reuse or recycling of the waste they generate, and they make an ugly mess of the place while rounding up only as much of their waste as they profitably can. Taxes pay for efforts outside the commercial sphere, but since design is geared toward sales only, simple-to-recycle containers and packaging are not forthcoming.

Now I need to revisit the work of two dear colleagues at the

Algalita Foundation—Marcus Eriksen and Anna Cummins—
who appeared in an earlier chapter but deserve more attention.
After learning of the research by Nikolai Maximenko and Peter
Niiler that identified potential "garbage patches" in the five major
high-pressure gyres of the world ocean, they started the 5 Gyres
Institute and launched research expeditions into the North and
South Atlantic as well as the Indian and South and North Pacific
oceans. Their research has established the presence, and measured
the quantity of, plastic in each of these oceans. Two people, organizing five expeditions on three boats over just two years, have
uncovered the "dirty little secret" hiding in plain sight on one
fourth of the planet's surface. Their motto is "Understanding
Plastic Pollution through Exploration, Education and Action."
Anna Cummins helped the Algalita Foundation organize the first
Youth Summit, which brought one hundred students from
around the world to Long Beach to brainstorm ways to deal with
plastic pollution. She and Marcus Eriksen continue to be a dynamic part of the movement to stop plastic pollution and can be
counted on to creatively battle the plastic plague for years to
come.

Somewhat to my surprise, the crusade against plastics pollution has gained a certain amount of cachet. Algalita was able to
get Ed Asner to MC our Coast and Ocean Connection Celebra-
tion at the Cabrillo Marine Aquarium. Graham Nash did a concert for us in Hermosa Beach, where we hosted a sort of public
teach-in on marine debris and displayed the *Fluke*, a raft Dr. Eriksen and the students at Environmental Charter High School
had made of PET bottles stuffed in an aluminum frame with
a sail ingeniously fashioned from recycled T-shirts and ropes
braided out of plastic shopping bags. It subsequently made a suc-

cessful journey from Santa Barbara to San Diego. Laurie David hosted my Technology Entertainment and Design (TED) presentation at her home for a group that included Ed Begley Jr. and Tony Haymet, the director of Scripps Institution of Oceanography, who was accompanied by the director of development at my alma mater, the University of California at San Diego. Dianna Cohen, artist of used plastic films and wife of singer Jackson Bowne, was in the audience and talked to me afterward about her desire to mount a campaign against plastic pollution. She followed through with the creation of the Plastic Pollution Coalition, which has coalesced the glitterati around the issue, and the coalition has become an international organization.

Single-use plastics are the major issue for many activists, since so few are recycled and so many harbor transmissible chemicals that end up in the ocean—and in us. The message is simple: If you are a manufacturer, redesign your products until they are nontoxic and easy to recycle. Hard to believe this would be impossible, but the objections are many. Here we see in bold relief the contradiction between marketing via brand recognition and simple design for enhanced recycling. The economics of unsustainable and unhealthful waste for profit is needlessly pitted against the economics of sustainable simplicity for reuse and recycling. The goal is not to limit creativity but to unleash it as the art of creating the truly "good" product, the one that will continue fulfilling human needs into the infinite future, that "plays the infinite game," as William McDonough says. The idea that we are bound to the current wasteful system because there can realistically be no other is to accept failure at the start of the game. Technology is poised to deliver sustainability but is corrupted by forces directing it toward the fulfillment of "special"—very "special"—interests that view humanity only as customers. The customers thus are made the agents of change, for without their support, those very special interests lose their profitability and with it the illusion of success, so why bother? Here lies the kernel of truth in the idea that organizing consumers to make truly smart purchases can save the planet.

CHAPTER 16

REFUSE

Without a variety of technologies, intelligently directed and
deployed, it will be impossible to strategically retreat from our
destructive engagement with the essential planetary systems.
The question isn't one of the use of technology or not, but
what kind of technology and at what risk. And to what end?

—DIANNE DUMANOSKI, *The End of the Long Summer: Why We*
Must Remake Our Civilization to Survive on a Volatile Earth

JULY 3, 2009. It is 0415 and I am underwater, in the middle of the
Pacific Ocean, hacking away at a ghost net that's wound itself
around *Alguita*'s propeller. I have the best tool money can buy
for this task: a sharp serrated bread knife. Breathing through a
snorkel, I work by the light of an underwater camera held by
"Scuba" Drew Wheeler, who treads water alongside me. We're
hoping the engine isn't damaged beyond repair. So are the other
four crew members, now fully awake and peering down from
the stern end of *Alguita*, ready to receive sections of net as I dive
down and cut them off. Though the *Alguita*'s a good sailor, we've
been periodically becalmed—not a good situation to be in, with

plenty of diesel fuel but a dead engine. And there's research yet to be done. It's only day 3 of our monthlong voyage to the International Date Line. Our mission: to sample a transect two-thirds of the way to Japan, where ghost net accumulations have been documented by NOAA scientists. We've already noticed a lot more "big stuff" roaming deep ocean waters: nets, crates, buoys, barrels, and strange items like toilet seats.

Just as I was falling asleep, back in my bunk, after an uneventful wee-hours trawl, I was awakened by a disturbing sound—silence. Joel Paschal, who crewed on the 2008 voyage, was on watch with Nicole, *Algalita*'s communications coordinator and first-time crew member. He met me on the bridge, where we attempted to restart the engine. We cringed at the metallic screech of steel on aluminum and headed for the engine room. For some reason the alternator was grinding into the hydraulic belt guard. We unbolted the belt guard and tried again. This time no screech, just a sigh before stalling when I shifted into gear. This alerted me to the need to grab mask, fins, and snorkel and check the propeller. Something's not letting the prop turn.

By now Scuba Drew is up and preparing his underwater camera and light assembly. We're about 250 miles north of Kauai, the northernmost of the populated Hawaiian Islands. The sea is fairly calm, with light winds whispering through the rigging in the pre-dawn darkness. But even in these mild conditions, removing a two-hundred-square-foot clump of green polyolefin trawl net from your catamaran's port propeller is not a safe or easy task, and wouldn't be even in daylight. You can get caught in the net and meet the drowning fate of so many marine mammals. The propellers are sharp enough to inflict a sizable gash as the ship bobs in the low swells. You hear stories. In fact, concerned friends seem to delight in sending items they come across, like the one about the tuna fishermen, one of whom wrote a letter to the U.S. Coast Guard that found its way to the Internet . . .

They were just a bit south of here, in the La Niña year of 2002, when one of two "partner boats" came to a grinding halt. Its propeller was frozen by coils of plastic net, rope, and banding straps bearing, they'd later learn, Korean lettering. The skipper nearly drowned in heavy April seas, working without benefit of diving gear to cut the propellers free. The other skipper left his boat to finish the nasty job as his buddy gasped for breath back on deck. In the letter to the Coast Guard, the second skipper described the episode and complained about debris "fouling" suffered with increasing frequency by the fishing fleet. He implored the Coast Guard to do something about it.

To someone who's been documenting the fouling by plastics of the North Pacific for more than a decade, this howl of outrage was less touching than ironic. Something along the lines of a coal-fired power plant operator complaining about mercury contamination in his swordfish. The UN estimates up to 10 percent of fishing nets and line are lost and abandoned each year, despite the raft of international laws. We've seen the damage it does. Paschal was part of a NOAA crew that helped remove a year's worth of fifty-plus tons of plastic fishing gear from the remote northern Hawaiian archipelago after certain havoc was wrought on fragile reefs and wildlife.

Irony truly abounds. Now when we put to sea to capture samples of plastic debris for later study, we find ourselves captured by plastic—it's the predator/prey role of plastic marine pollution at the human level. In all its many forms, it lasts far longer in the ocean than on land. We've found that this miracle of man's ingenuity—this material that forestalls natural decay yet puts nature at risk—has itself become an element of the marine environment. We've concluded that this is a crime against nature on a massive scale. And it's made the north-central Pacific too dangerous a place for investigations conducted on a fifty-by-twenty-five-foot catamaran. Assuring the safety of my crews and the integrity of my vessel has become too great a challenge for this sea captain. Once upon a time we worried about our boats getting tangled up in coastal kelp beds. Now the kelp itself—what remains of it—is choked with plastic trash, and there are faux forests of plastic kelp in the middle of the ocean.

We're lucky. With the last of the net removed, we have a closer look at the engine. We see that the net wrenched the driveshaft to a halt with enough force to thrust a five-hundred-pound motor a full inch forward on its mounts. But the only visible sign of damage is the gouge on the housing for hydraulic belts—purely cosmetic. With the transmission in neutral, I gingerly turn the ignition key, then pop the engine into gear. It's good.

The propeller net scare occurs on the second leg of a three-part decadal anniversary voyage. That leg is a northwesterly hop from Honolulu to the International Date Line. The final leg will be a reenactment of the 1999 sample protocol unfolding on the voyage back to Alamitos Bay. For this last segment I have a fresh crew that includes our prized "boat monkey," Jeff Ernst, as first mate; Algalita's dedicated marine biologist, Gwen Lattin; Lindsay Hoshaw, a recent journalism grad who will be the first to tweet from *Alguita*; Bonnie Monteleone, a University of North Carolina grad student; and Bill Cooper, the director of the Urban Water Research Center at UC Irvine. We're running a month later than the '99 study, with trawls starting in late September and executed in reverse order, starting on the western edge of our sample design. We find what a difference a decade and a month can make. The sea state—measured by the Beaufort scale on

which 1 is flat and 12 is "perfect storm"— builds over the course
of our sampling from 3 to 6, with fifteen-to-twenty-five-knot
winds and three-to-four-foot swells—middling conditions at
best for surface trawls. We've learned particulate plastics tumble downward through the water column in active seas, with
smaller particles slow to resurface. In 1999 we had perfect conditions, with five-to-ten-knot winds and one-to-two-foot swells.
In 1999, we never filled the cod end of the manta net with the
plastic-plankton mix. On this trip, goopy plankton loads the collection bag on several trawls, suggesting rougher seas may be
churning up more nutrients. It takes three days to complete the
ten-year retrospective study.

A week later, we cruise into Long Beach Harbor. It's Wednes-
day, October 7, 2009, and we're glad to be home after a four-
month journey—*Alguita*'s longest yet. We enjoy a boisterous
reception on the dock and nearby Algalita Marine Research Foundation headquarters, then run the samples over to the Sea Lab.
Analysis protocols are the same as in 1999, except this time the
plankton samples are lumped, not individually parsed for number and species. It's a high-priority job, completed in comparatively short order by Gwen Lattin, Ann Zellers (who processed
our first samples ten years earlier), and several lab volunteers.
With so much plankton, one sample can take several weeks.

The results intrigue. In 1999, we captured a total of twenty-
seven thousand pieces of plastic. In 2009, we collect "only"
twenty-three thousand. What's somewhat confounding is that the
same eleven trawls on the winter 2008 voyage—the lanternfish
study—produced sixty-two thousand plastic bits. Studies such as
ours are typically "corrected" for variables, and the agitated 2009
sea state would be such a variable. But another measurement,
weight, further complicates matters. In 1999, total dry weight of

particulate plastics was 424 grams. In 2008, it was 669 grams. In 2009, it was a "whopping" 881 grams, nearly two pounds, and nearly double the 1999 weight. This finding affirms our perception about a growing abundance in the gyre of macroplastic debris. But a couple of fairly large pieces in a travel sample can skew results, so the disparity between 2008 and 2009 seems less consequential than the fact that both '08 and '09 weighed substantially more than '99's sample. We observed more big objects in 2009, the entangling ghost net being just one of many. It seems plausible that earlier generations of plastic trash have "gone nano" while plenty of new trash is midway through its journey to nothingness.

I should take some time to deal with the issue of the plastic-to-plankton ratio. Of all the ways we measure plastic pollution of the marine environment, this has caused the most controversy. The variability of plankton on the ocean surface is tremendous, but so, as we have seen, is the variability of plastic. In 1999, the ratio of dry weight of plastic to plankton was six to one. In 2008, it was forty-six to one, and in 2009, it was twenty-six to one, which is surprisingly high, considering these were the most plankton-choked samples we'd ever taken. Remember, we compare plankton biomass to plastic mass as a measure of potential harm, because comingled plastic and plankton form an edible matrix for indiscriminate surface feeders, from salps on up to albatross, sea turtles, and baleen whales. On many night dives, when the underwater world teems with fantastical undulating life-forms few ever see, I have closely watched the behavior of lanternfish surface feeding. Once these fish make their nightly ascent, perhaps from a mile deep, it is not their way to leisurely pick and choose from among available zooplankton. They feed frantically, indiscriminately darting, scrambling, and rapid-fire nibbling. In our 2008 fish study, we'd found that 35 percent of them had ingested plastic of all colors, apparently not selecting for color in the darkness. Like so many other marine creatures, they're being duped by plastic bits, which so closely resemble natural food in size, shape, texture, and passivity. If the ratio of plastic to plankton increases—in 1999 we had a sample in which plastic pieces actually outnumbered individual plankton—then it follows that ingestion of plastic by planktivores will also increase. And deeper into the food web plastic penetrates.

But we have been called out, both by fellow researchers and spokespeople for the plastics industry. This is what they say: your statistics are shocking and you use them to bait the media. You make it sound like the entire ocean is paved with plastic and we're all doomed. Then there's the false equivalency argument. It goes like this: you can't extrapolate from your samples because the plankton population is variable throughout the marine environment. Your research doesn't allow that there are places where there will be more plankton than plastic.

The first objection is easily addressed. The media naturally seize on the most attention-grabbing aspect of any story. It's their business, and to stay in business they have to pique curiosity. Who can change that? We've done our share of cringing when we see the Garbage Patch described as a "swirling continent of plastic" and such. But is the risk of distortion a reason to avoid the media when you have an important message? We've been circumspect in our messaging, and we've publically corrected a number of misconceptions. But we can't control media interpretations of our story, and we dare to hope that some good comes of them. The second objection is answered by the lanternfish. Sure, we calculate the density of plastic and its abundance. But what really matters is the evidence we've gathered that shows plastic insinuating itself into the marine food web. If we don't characterize the available food and compare it to plastic debris, we're ignoring the most crucial issue.

ALGUITA HAS TRAVERSED one hundred thousand sea miles in fifteen years of oceanographic research. She is a modern, well-equipped research vessel. And she can really sail! Being technologically sophisticated, she has several high-maintenance systems, but all oceangoing boats are high maintenance. They sail in what amounts to a caustic, non-compressible saltwater bath—and they occasionally collide with fixed and floating objects, some seen, some not. (Stephen Colbert nailed it when he said the ocean "has gone from smooth to extra chunky" during my appearance on *The Colbert Report*.) *Alguita*'s maintenance requirements are orders of magnitude greater than those of a modern car. I like to say that all boats are FORDs, an acronym from the Model T era standing for Fix Or Repair Daily. As her engineer as well as captain, I am kept busy: maintaining her mechanical systems, repairing her equipment, setting and adjusting her sails, leading research efforts, even doing most of the cooking. It's a lot of work, for which I am rewarded with "the poise and peace of labor," as Zane Grey put it.

The enormous ocean still awaits pioneering research that *Alguita* is uniquely suited to do, especially in the mid-latitudes, where the gyres are. I'm working on some modifications that will make her more "debris-resistant." Where *Alguita* really shines is far out at sea, in the global commons that no town, city, state, or nation is obliged to monitor or protect. Now I just have to figure out how to keep her props clear of debris and reinforce the above-water plating between her two hulls, which is getting holed and split by massive wave impacts and collisions at night with hard plastic floats.

Sailing over the open ocean in the temperate latitudes, I appreciate the vitality of the sea's embrace. As originator of life, as regulator of climates, and dominant feature of our "biosphere," the ocean deserves to be accorded rights, to have standing in the legal sense. We should know what's going on with her, listen to her complaints, and protect her. After all, we describe her in human terms: she has a circulation and delicately balanced systems. She has moods, mysteries, and charm. She's very deep. She can be healthy or sick, and she is clever, evolving in ways so as to maintain balance or, technically, homeostasis. I hereby present an amicus brief on her behalf as her voluntary public defender. The sea has a right to be rid of the toxic waste of civilization. We cannot sail over her and remove it any more than we can fly around the sky and remove chlorofluorocarbons that caused a hole to form in the ozone layer. What we can do is stop the flow of trash that daily adds to her burden, then give her assimilation capacity a chance to degrade our plastic excrement to insignificance. We would record a gradual diminution similar to that of the ozone hole since CFCs were banned.

But synthetic polymers are different. In earth time, they are brand new, and as yet no efficient assimilation mechanisms have evolved. If, in ecosystems, everything is used, how will earth's ecosystems use plastic waste? Is killing millions of animals a "use"? Is the toxifying of our bodies, and those of generations to come, a mere "side effect," a small price to pay for the privilege of participating in the benefits of the Plastic Age? Groups and individuals are taking notice and taking action. Relevant laws are in place, but as we've seen, polluters are often shielded by a different set of laws, creating an impasse we can ill afford. We need the conviction, and the clout that comes from organized effort, to resolve cross-purposed legislation.

It lives in me . . . the experience, now a memory, of a plastic-free ocean teeming with an abundance of mature marine species. I grew up as we left the pre-plastics era, and I experienced the awesome, inspiring ocean of Jacques Cousteau. My generation now leaves the somber veil of plastic pollution as our legacy for many future generations to deal with. This sad reality is scientifically demonstrable. That it is wrong to leave such a legacy we call a value judgment. But can the two be separated? Doesn't the fact that our "plastic footprint" is killing millions of marine animals contain the value judgment that it should cease? For those who may see the increase in bacterial and sessile organisms on ocean plastics as increased biomass and new habitat, somehow justifying the animals starved, strangled, and poisoned by plastic waste: Get real! Biodiversity is decreased by exotic species colonizing faraway habitats via plastic transport. Plastic's power to create fascinating technological possibilities is now morphing into a power to disgust and alarm us with its unintended consequences. Will we evolve a response faster than the albatross and reject what harms us and our environment?

What I see happening to the ocean breaks my heart while forcing open a yet-to-be-named field of science. (Somehow "marine garbology" lacks the gravitas appropriate to a science.) Can anyone dispute that we live in the Plastic Age when two years of plastic production equals the weight of every man, woman, and child on earth? When it takes the shape of anything we want and surrounds us at work and play? I call plastic "the solid phase of petroleum," and think of it as polluting like an incredibly dispersed oil spill, an oil spill that lasts centuries and mimics food while sponging up toxics.

Curtis Ebbesmeyer asks, "What are ocean plastics' fundamental properties?" He wants to know how much of the ocean's surface is covered, and the quantities in all the places plastic debris accumulates. The nations of the world, which are all contributing to the problem, should embark on a program to find the answers. *Alguita*'s trained citizen-scientists, deploying our "traveling trawls" will help in this effort. But in the meantime, how about this idea: let's study plastic's properties and effects *while* we cut out throwaway plastics and stop urban runoff. Too often the need for deeper understanding is used to stall remediation. The reason I founded the Algalita Marine Research Foundation was to shorten the distance between research and restoration of the marine environment. The "spill, study, and stall" crowd—advocating the hundred-year-old petroleum industry's strategy—demands a science of valueless facts to provide a "complete" mechanistic understanding of a problem before embarking on any solution. Endless, purposeful delay pending perfect "sound science" enforces a form of intellectual sadomasochism driven by the need to preserve profits, not benefit society. Because valueless, mechanistic understanding enables technological mastery of materials and processes, it "delivers the goods." But it is irrational because it fosters those all-too-ubiquitous side effects, the euphemistically termed "unintended consequences," actually predictable but unwanted consequences, which are assiduously downplayed as much as possible to ensure initial, quick profitability.

Of course we want the best "bang" for the restoration "buck," but the finest minds on the most influential committees haven't found a way to rein in the plastic monster. Twenty-five years after the first of five International Marine Debris conferences, the oceans' plastic load continues to increase. We can't economically reverse the trend, since our "strategic plan," as the dominant species globalizing a historically derived brand of civilization, is to hang on tight and "blow out all the stops" on our way to higher levels of growth—in production and consumption—the sooner to reach a global democracy of work and consumption modeled after the rich countries.

We are faced with a fundamental contradiction. The economic system that brought us fabulous wealth and unprecedented growth can't give us, as a basic return on our investment of lives, labor, and loyalty, a healthy planet. The ark that has carried us and the animals this far is foundering, and it can't take us the rest of the way—the rest of the way to technically possible safe harbors for all human and animal families, where real abundance and individual potential are realized and we all thrive in biodiversity.

For some, the shame and outrage that come with seeing our waste plastics harming innocent creatures and trashing beautiful lonely places helps create this type of "political will." I certainly seem to be delivering a message people want to hear: from *Pravda* to Fox News; from the *Late Show* to *The Colbert Report*; from the morning and evening news on CBS, NBC, and ABC to their counterparts in the Netherlands, Australia, Italy, and France; from the TED conferences to meetings of the European Commission, the World Federation of Scientists, and the Club of Rome; from Plastics News to the Campaign Against the Plastic Plague; from the assembly hall at my old elementary school to a packed house of more than six hundred on a cold and rainy Monday night in Moscow, Idaho; from a Rotary breakfast at Friday Harbor to a Surfrider Foundation "Rise Above Plastics" conference in Honolulu, with a film shoot by Korean TV at

Point Panic with Diamond Head in the background—audiences have heard me say our plastic footprint may be causing more immediate harm to sea creatures than our "carbon footprint." And they have approached me afterward, moved, angry, and motivated to change their personal habits and spread the word.

The title of this chapter is "Refuse." What is meant by this fourth R? Should it come first, as in "Refuse," "Reduce," "Reuse," "Recycle"? I understand it in a larger context. To me it means taking part in what one of my UCSD professors called "The Great Refusal."

Indulge this sailor as he revisits his roots and lays down his "rap." The refusal to be an uncritical part of the economy, and the refusal to blindly support the winners on the winning team, no matter the "side effects," have guided my actions since I dropped out of college in 1967. The refusal is never a clean break, for we must always sail within the "trade winds of our times," but much can be avoided, in spite of steep penalties for not "playing the game." Today the movement to support local economies by purchasing locally produced goods has the added benefit of eliminating the packaging of globalization, which accounts for much if not most of the ocean's plastic pollution. The time seems ripe for a strategic withdrawal from global commerce in order to "tool up" for major change. World trade is a reality, but the more we trade locally, the less throwaway packaging we need. A practical way to begin changing the pollutingly successful yet blindly moribund throwaway economy is through local self-reliance.

In the United States the Institute for Local Self-Reliance "works with citizens, activists, policymakers and entrepreneurs to design systems, policies and enterprises that meet local or regional needs; to maximize human, material, natural and financial resources; and to ensure that the benefits of these systems and resources accrue to all local citizens." This movement is already happening in enlightened pockets of the world, and it's not going unnoticed. A February 2011 edition of the international trade e-newsletter Foodproductiondaily.com contains this startling suggestion: "The food industry should not rage against the idea of professionalised local food systems, nor unleash its lobbying force to uproot them before their green shoots can reach maturity. Rather, it should explore ways to benefit from local foods and, in turn, foster their development." The title of this piece: "Why Local Food Systems Are an Opportunity for Industry." The grip of established brands and markets will be hard to shake off, and co-option by big business is always a threat, but local ownership of businesses that can "deliver the goods" will create employment and catalyze change.

The science and technology that will liberate us from pollution and meaningless labor are available. The resources and raw materials begin at the local recycling center, better named "resource recovery" center. The employment opportunities we create will let us live our lives and do our work without sacrificing the beauty of our natural surroundings as our plastic and carbon "footprints" fade. We can discover and characterize life's true needs as scientifically as we want, let them develop naturally or carefully produce them, and liberate ourselves and the natural world without succumbing to the perverting pressures that reward rapid growth and profitability. We can do this if we own and skillfully employ the instruments of science and industrial production on a local scale, while respectfully and intelligently managing our local ecosystems. Small shops and farms can do big things. Schoolkids race electric vehicles they make on campus. Let's sponsor a contest to make the best electric delivery vehicle for local organic produce.

One of the first focuses of the Algalita Foundation was on-site sewage treatment in response to septic tank pollution of the famous Malibu surfbreak. Our president emeritus, Bill Wilson, is an expert in this field. He refers to himself as a used-water salesman. Localized sewage treatment, in a series of underground plastic settling tanks that distribute the treated sewage under mulch cover to landscape plants, was approved by the City of Malibu's Sanitation Department, a first for a residence in Southern California. You see, we can develop the basics and integrate them into *Cradle to Cradle*'s "Infinite Game." And as our success in producing the necessities of life without harm increases, we lay the groundwork for eliminating the plastic plague.

I believe a shift in economic focus will be a prerequisite for ending plastic pollution of the world ocean.

In the United States at the beginning of the second decade of the new millennium, we find ourselves with 10 million workers unemployed, yet supermarket shelves are full, Internet orders for anything you want arrive in a couple of days, and you can get your dog's nails done while you choose from an assortment of treats for your best friend, based on the dog's age and physical condition. What are the 10 million supposed to do? Enlarge the services sector? We've got everything we need and much, much more available. Yet the mantra is "Innovate!" And "Export!" (Something other than jobs, we hope.) Easier said than done. We're told small business is the backbone of the economy—but that's what they said before factories were shuttered and whole cities that grew up around them fell into decay. In any case, as we've seen, growth too often means millions of new "must-have" short-lived products—pollutants, really—soon destined for the landfill, or the ocean. Other countries resist importation of goods they want to make for themselves. Modern statecraft is increasingly the art of getting other countries to buy your country's goods, even if your cheap, subsidized exports crush their local economies. Few benefit but they benefit hugely.

Perhaps there's irony here too. We seek self-reliance, and places to incubate it—a conservative value if ever there was one. Even some mainstream schools are adjusting to what appears to be a new reality, one in which good jobs aren't there for the plucking and equipping graduates with market-ready skills is sound pedagogy. We must insist that the "work/study" curriculum move toward an economy that makes sense for everyone and respects the natural environment that supports it.

The generation that will stop plastic pollution in its tracks will be the generation that has unplugged itself from the economy of constant input and output of junk. Pointless competition and unthinking consumption will be replaced by mindful acquisition of goods that are needed and will last and will be fixable if they break. And when they have served their purpose, they can be refashioned as something worthwhile. These hardworking, creative, peaceful warriors will reject an unsatisfying wealth of "new" commodities and find true wealth in their pursuit of beautiful, productive lives that waste not. They will reject the lifestyle of underappreciation and overconsumption of shoddy goods for one that respects aesthetically organized labor and the wonderfully useful products it creates. Their heroes will be the producers and re-producers of authentic necessities that bring health and happiness. Knowledge of technology will be shared as easily among them as knowledge of organic farming and gardening is shared among its practitioners today. They will not fear Mother Nature but see in her another being, striving as they are to perfect the loop of life. They are not the "now" generation, seeking to be "present" in a polluted metropolis. Nor are they the "next" generation in an incremental movement toward an "improved" polluted planet. They will be the generation, perhaps a generation born of crisis, that breaks with the past as with an overbearing parent whose oppressive rules and antiquated ways are now obsolete. The novelty they seek will not be that of a new distraction or pseudoconvenience but the novelty of creating something truly liberating, for themselves and the planet.

Adam Smith said there is an "invisible hand" that regulates the economy through economic self-interest. That invisible hand is now being overshadowed by the visible hand of Mother Nature herself—her self-interest and her non-negotiable limitations. She is the beginning and the end of our economic process, and we must know and respect her "ecosystem services." Let's combine these ideas and evaluate products so when they go to market they can be labeled with the characteristics that matter:

1. Closed loop recyclability index: How easy is this product to recycle?
2. Extended replacement time rating: How long will this product last?
3. Reduced maintenance time rating: Is the product maintenance-free? (Here is where plastic belongs: in long-lasting, maintenance-free products.)
4. Potential number of products replaced or made obsolete: Does this product eliminate the need for a lot of other products?
5. Raw material extraction stress index: Is the product 100 percent post-consumer material?
6. Nontoxic status: Are the components benign from a biological perspective?

the "next" generation in an incremental movement toward an "improved" polluted planet.

And why not scientifically evaluate products for those oft-expressed values of freedom and individual liberty, now ritually invoked to justify expansion of the status quo? How about the "product set the world free index" and the "product human liberation index"? (How many of the above qualities do they have?) These values barely exist in the toxic political lexicon of today's national leaders. Let's give these concepts real meaning and use them to regulate the production of goods! The longer we wait to get serious about what these terms really mean, and then to internalize them as ways of being, the greater the risk we impose on posterity. We need to create the space where the future can blossom, where "The realm of mind achieves in freedom what the realm of nature achieves in blind necessity—the fulfillment of the potentialities inherent in reality."*

The seductive idea that the more we consume, the better off we'll be has timed out, and the Plastic Ocean is one of many witnesses to this fact. Who would have thought the ocean itself would be such an effective advocate for zero waste? Who would have thought that broken bits of plastic bobbing on the deep, wide ocean would catalyze a significant political movement to change the way goods are produced and consumed? We have no trouble imagining a zero-waste closed loop in agriculture, with plant material and food waste composted to make productive soil for the next crop. It has worked just fine for millennia. The time has come to close the loop with goods as well as commodities. It won't be as easy as in agriculture, because human ingenuity must substitute for compost organisms in rendering the next cycle's raw materials fit for production, but we have the smarts, the

---

* Herbert Marcuse, *Reason and Revolution*.

know-how, and the imperative. The ocean planet will thank you if you help end its plastic plague.

I am a patient man, and I have learned the art of seeing. I saw small scraps of plastic in the middle of the ocean that others missed. I understand process and have learned by doing. And I know how a few well-placed nudges can alter a course, the way a slight tug on a ship's wheel will point you toward an entirely different destination.

AFTERWORD

THE HARDCOVER EDITION of *Plastic Ocean* was in final edits on March 11, 2011, the day northeast Japan was ravaged by earthquake and tsunami. Watching TV images of coastal towns being swept out to sea stirred feelings of dread, not only about the dire human toll, but about impacts this massive unleashing of debris would have on North Pacific ecosystems. We slid a few allusions into the final text, but at that early stage, as the tragedy still unspooled, there was little to say that wasn't speculative. Even now, though we know more, hard data are yet to be collected, though observations are accumulating.

Publication of the paperback edition affords a welcome chance to note what's been learned thus far about this unfolding story that will be written over decades. Japanese officials estimate that the tsunami waves raked 5 million tons of debris into the Pacific, of which a probable 70 percent would have sunk offshore. A year's worth of ocean dumping in one day. The 1.5

million buoyant tons now traversing the North Pacific are equivalent to a bridge of compact cars, three abreast, spanning the entire North Pacific. How much of it is plastic is anyone's guess. Initially the debris clumped, but now it's dispersed and undetectable by satellite camera. Some is stranding along northerly North American shores (California will be largely spared), and soon Hawaiian beaches and atolls will be littered. Oceanographer Nikolai Maximenko predicts that as much as 95 percent of this rubble will remain captive within the great clockwise current circling the North Pacific, from there to be swirled into the eastern and western accumulation zones (aka garbage patches).

As we write, a 5 Gyres Institute expedition, co-sponsored by the Algalita Marine Research Foundation, has just completed a sampling study of the Northwest Pacific Gyre, home of the so-called Western Garbage Patch between Japan and Hawaii. The research crew, headed by Dr. Marcus Eriksen, Algalita's longtime associate and 5 Gyres cofounder, aimed for a snapshot reading of tsunami debris in this predicted hotspot but had to conclude that its copious plastic flotsam, equivalent on first look to what's bobbing around in the Eastern Gyre, could not be definitively linked to the tsunami. Lab analysis of samples taken here will enrich the skimpy database for this rarely studied area of the Western Pacific. On the other hand, post-tsunami research in the well-studied Eastern Garbage Patch should yield the best data yet about the way marine debris moves through ocean waters, how it degrades and how it affects marine creatures in this gyre. We'll have a debris forensics advantage knowing the source of new material traceable to Japan's tsunami.

With deeper understanding will come refinement of computer models designed to track and predict debris impact. For example, the premier debris-drift paradigm, developed by Nikolai Maxi-

menko and Jan Hafner of the International Pacific Research
Center at the University of Hawai'i, initially predicted North
American landfall of tsunami wreckage between March 2013 and
March 2014. But in April 2012, after just one year, a Japanese
ghost ship set loose by the tsunami arrived in Alaskan waters
well ahead of schedule and was sunk by the U.S. Coast Guard.
(The Coast Guard and U.S. Navy are on the lookout for ship-
ping-lane hazards of this kind.)

Finding beach debris with Japanese markings is commonplace
along shores in the Northwest, making it tricky to link individual items to the tsunami. But shortly after the ghost ship turned
up, a beachcomber on Graham Island, British Columbia, found
a crate containing a set of golf clubs and a Harley-Davidson mo-
torcycle registered in the hard-hit Miyagi Prefecture. The crate
was traced to the owner, who had lost three family members in
the disaster. (In a gracious gesture, Harley volunteered to restore
and return the bike to its owner free of charge.). The Hafner–
Maximenko model also predicted that debris would reach Midway Island in the winter of 2011–2012. I checked with Cynthia
Vanderlip, wildlife manager at Kure Atoll, some fifty-six miles
northwest of Midway. She said the debris season was lighter than
usual, a sign that tsunami flotsam was churning eastward through
waters farther north of the Hawaiian archipelago than predicted.

As a NOAA official noted, "We're working on updating the
model." Indeed, it would appear that the debris *is* the model,
with where it will go well enough known and when it will arrive
the greater mystery. NOAA is already conducting volunteer-
training sessions in Oregon coastal towns, imparting protocols
for tsunami debris recovery. And it is reasonable to surmise that
things now landing on North American shores are out front of
the bulk of the debris. Superbuoyant debris items, things like

inflated soccer balls and lightbulbs, have an obvious speed advantage. Riding high on the ocean surface, these floaters catch brisk westerlies and skim ahead of the slower current beneath them.

It's no easier to predict with precision the ecosystem damage sure to unfold as debris fans, strands, and accumulates. New research that updates the pre-tsunami baseline is well worth noting here, starting with a study shared by Miriam Goldstein, a doctoral candidate at Scripps Institution of Oceanography. She led a meta-analysis of existing plastic debris data, including Algalita's, and determined that plastic in the ocean has increased a hundredfold over the past forty years. It's a result depressingly consistent with my 2009 trawls, many of which seemed improbably plastics-laden. Another recent study, led by a University of Delaware team concluded that marine plastics could be twenty-seven times more abundant than currently estimated. The researchers found strong evidence that ocean turbulence is churning vast clouds of plastic bits deeper down in the water column than previously thought. Thus they escape detection, by eluding surface trawls.

Goldstein also shared a sobering discovery connected to her specialty, the study of marine organisms as they interact with man-made objects adrift in the ocean. To wit: The marine insect *Halobates sericeus*, a saltwater skimmer native to the North Pacific Gyre, has embarked on an unprecedented egg-laying spree. This bug deposits its eggs on ocean debris, once only natural materials like driftwood and pumice. With the rise of waterborne plastics, there's much more debris, hence a frenzy of egg-laying. This could be a good thing for the insect's main predator, crabs, but it puts the larger marine food web at risk. *H. sericeus* dines primarily on tiny animals like zooplankton and fish eggs. If these species take a big hit, so will lantern fish and other planktivores.

It's a perilous food-chain disruption that will only be amplified by tsunami debris.

This sort of thing is just the tip of the iceberg. A point I now make in my presentations is that the plastic in the gyre is creating a new coastal habitat. We've come upon mid-ocean equivalents of tide pools replete with sea anemones, oysters, mussels, barnacles, and worms clinging to stray buoys, net clumps, and other macro-debris. We've found crabs scooting around in these colonies and seen reef fish darting among new ecosystem features. On Marcus Eriksen's blog for the Western Gyre voyage, he reported finding a five-hundred-pound ball of netting and debris north of the Marshall Islands. He wrote: "Fish are everywhere—mahimahi, amberjack, triggerfish—circle beneath the net ball. . . . We haul it above the deck to shake it out. More fish, a goby, 5 frog fish, hundreds of crabs, a shrimp, worms, nudibranchs, anemones—Hank Carson from U. of Hawaii collects 26 species in all." It's an evolved mid-ocean ecosystem. I've just returned from Japan, where I met with these voyagers and sent them off their next leg: an ambitious expedition that will trace the path of tsunami debris, gathering samples along the way. The mission includes the study of debris "fouling," the term used to describe biological growth on waterborne detritus. We'd like to know how these micro-islands of life evolve.

You might ask, as some have, What's the big problem with a mid-ocean population boom? Life, after all, is a good thing. But life is not such a good thing in places where nature didn't intend it to be. There are cases on record of organism-encrusted flotsam reaching distant shorelines and seeding balanced ecosystems with aggressive invaders. And let's not forget that these organisms are rooting themselves on a material, plastic, that we've learned not only sponges up long-lived toxic chemicals from sur-rounding waters, but also harbors and releases bioactive chemi-cals used in their making. More plastic debris equals greater potential toxification of the food chain, starting with entry-level denizens and ending with us.

We'd also like to delve into recent findings about this toxicity. The plastic-related compounds covered in Chapter 13, "Bad Chemistry," the sturdy, persistent halogens, are mostly associ-ated with proven risk. Some, like PCBs and DDT, have been banned and are dissipating, though they still linger in the envi-ronment. Activists like Berkeley-based Arlene Blum of the Green Science Policy Institute appear to be breaking through with warnings about brominated flame retardants. A May 2012 inves-tigative series published in the *Chicago Tribune*, considered a game-changer, prompted Illinois senator Dick Durbin to call for an end to their pervasive use in electronics and home furnishings.

But among other chemicals dubiously associated with plastic, no two loom larger than bisphenol A (BPA), the core ingredient of hard polycarbonate plastic as well as epoxy adhesives, and phthalates, which impart flexibility to vinyl and other materials. They are noted in "Bad Chemistry," but not thoroughly ex-plored, since, in some quarters, their health effects are considered too "speculative," the science too unsettled. Indeed, the same verbiage used to cast doubt on climate change is routinely ap-plied to these chemicals. And with both, avalanches of red-flag science seem to be doing little to speed the glacial rate of policy changes that would undoubtedly make us safer. Why? It's very simple and no secret: Science often finds itself at odds with com-mercial interests that have much to lose. It turns out that science can be employed as easily to discredit as to prove. Results depend on who's doing it, who's paying for it, and the methods used, because protocols lack standardization. And the stakes are very high indeed.

Take BPA. It's an $8 billion industry. More than 6 billion pounds of BPA are produced annually by Bayer, Dow, and several lesser-known chemical companies. A quarter of it becomes epoxy lining for metal cans. Much of the rest is used to make polycarbonate plastic (for eyeglass lenses, five-gallon water bottles, DVDs and CDs, toys, and headlight covers, among other things), thermal receipts and carbonless copy paper, and dental sealants. (A less familiar application is in fabric finishing). Trade-secret laws very likely conceal its presence in many other products. The EPA estimates that a million pounds a year leach into the environment. The result is that about 95 percent of Americans test positive for BPA exposure, and for phthalates as well— even people with "green" lifestyles. And yet its industry boosters claim BPA is quickly and harmlessly metabolized and excreted by the human body.

To explain this stunning level of contamination, you need only consider our plasticized daily lives. We sleep on polyurethane "memory foam" mattresses and pillows stuffed with polyester fiber or polystyrene foam beads, covered by fleece blankets spun from recycled PET bottles. In the bathroom, the (acrylic) tub, toilet seat, shower curtain, comb, brush, toothbrush, product and pill bottles, even the pills with their polymer "enteric" coatings—plastic. Almost everything baby. Sport shoes. Synthetic fabrics are woven polymer fibers. Eyeglass frames are mostly "nice" plastics like nylon and zylonite (cellulose acetate, like cigarette filters and film stock), but lenses are typically polycarbonate. Wall paint, furniture coverings, non-natural carpet and curtain fibers, seat cushions, vinyl flooring, window frames, and house siding—all synthetic polymers. Chewing-gum base is styrene butadiene (think car tires), polyvinyl acetate, or polyethylene. Gum makers aren't keen to share particulars. Melmac dinnerware, children's sippy cups. Microwave, fridge, and dishwasher, all lined with plastic, typically hard polystyrene. (Styrene, also the basic ingredient of Styrofoam, has been named a "likely carcinogen" by the National Toxicology Program, over industry opposition.) Food and beverage packaging, fast-food accoutrements, bags, bags, bags. Electronics—phones, computers, TV remotes—handled constantly. (Experts say chemical air pollution is worse indoors than out in most places.) Car interiors.

Then there's invisible stuff. Plastics lurk inside our walls as insulation, carry tap water to our faucets, and provide substrates—typically PVC gutter—for hydroponically grown vegetables and aquaponically raised seafood. Municipal water systems extract sediment from drinking water with a polymer flocculent. Epoxy sealants line water storage tanks and pipes. A study in the 1990s found that fuels and pesticides, ground contaminants, were permeating buried polyethylene water system pipes and entering the water supply. Municipal water systems test for biologics, metals, minerals, and turbidity, but not always for contaminating chemicals.

Experts view these relentless, endless, and countless little exposures—from food, the air that we breathe, and skin contact—with growing concern about human health consequences. And what a mushrooming body of science is telling us is that many of these chemicals may actually be changing us through a process known as endocrine disruption. Endocrine-disrupting compounds (EDCs) have turned out to be as diverse as they are inescapable in almost every aspect of almost everyone's life. By definition they are substances, both man-made and natural, that alter biological processes by mimicking natural hormones, most disturbingly during fetal development. In popular parlance, EDCs are called "gender-benders," because they tend to be estrogenic and to feminize males, but that term is limiting. Health effects from EDCs can ripple throughout a biological system. Maladies thought to be linked to EDCs include a number of diseases and disorders that seem freshly emergent or newly ubiquitous: obesity, type 2 diabetes, autism, ADHD, thyroid dysfunction, asthma and other autoimmune disorders, childhood cancers, breast cancer, infertility, preterm birth, and miscarriage. EDCs have produced many of these health problems in lab animals, but human studies pose obvious challenges. Unlike lab rats, we're exposed to such a welter of contaminants each day that it's well nigh impossible to thread the needle of specific causes and effects.

The list of EDCs includes the halogen compounds, but also fungicides and antimicrobials, several categories of pesticides, mercury, cadmium, lead, soy (the strongest of plant-sourced estrogens), and a number of industrial hydrocarbon-based chemicals including BPA, a mild estrogen, and phthalates, which block the male hormone androgen.

It's not so surprising that a chemical designed to kill bugs could also prove harmful to higher-order animals. But it's a little surprising that chemicals that catalyze, harden, soften, strengthen, flameproof, waterproof, color, or impart any number of performance characteristics to synthetic polymers could turn out to be not only bioactive, but bioactive in ways that may permanently alter an organism's essence. Thousands of animal studies have shown this to be the case. The National Institutes of Health has an online aggregator for peer-reviewed publications in the life sciences. If you search "BPA," 2,962 individual studies come up. Among findings reported in recent months:

- Prepubertal female mice exposed to BPA experience early puberty and impaired fertility. In another mouse study, BPA altered mammary gland tissue and altered breast milk.
- Pregnant rhesus monkeys dosed with BPA levels comparable to those of the average American produced female offspring with pre-cancerous breast tissue.
- Male mice dosed with BPA around the time of birth exhibited impaired testicular function and lower testosterone levels. That is, BPA lowered sperm count and motility.
- Adult male mice treated short-term with BPA developed a propensity to gain weight.

At a 2006 World Federation of Scientists conference held in Sicily, the Italian scientist Paola Palanza drew a collective gulp with data showing that BPA-exposed mouse mothers avoided nest time with their babies while similarly exposed males were more mommy-like. I was one of those who gulped.

In a lab setting, each tiny aspect of an experimental animal's life is controlled, but the methods used to test them often vary. For example, dose matters—its size, duration, and delivery method. Toxicologists tend to adhere to the Paracelsian credo that "the dose makes the poison." Translation: A big dose might kill, but a small one will do little harm. Yet endocrinologists and cell biologists inhabit a different paradigm, where infinitesimal doses, parts per billion, can get big results. Also key is the timing of the dose, especially when the study subjects are unborn babies. Placental incursions by EDCs can alter a genetic blueprint if they occur during specific developmental moments as the baby undergoes cellular construction.

No less crucial is the method of dosing. Oral doses are metabolized through the liver and have a milder effect than injected doses. Thus, same-sized doses can produce very different results. Duration also matters. If you want to duplicate the effect of everyday "background" EDC exposures, dosing needs to be low and continuous. Yet some researchers deliver one big dose. Finally, specially bred rodents are considered reliable stand-ins for human study subjects, *unless* the results happen to displease an industry group.

With humans as subjects, studies are even trickier, since virtually every American already carries traces of BPA and phthalates—a fact gleaned from an ongoing testing program conducted by the federal Centers for Disease Control (CDC). The CDC regularly surveys about 5,000 geographically and demographically diverse Americans, compiling health data and testing for more than two hundred metals, minerals, and chemicals, called the body burden. Most Americans harbor well over one hundred chemicals, albeit in small traces. Breast milk is thus "fortified." Children have proportionally higher burdens and are less able than adults to self-detox. The data provide fodder for research, but the CDC pointedly cautions that chemical positives don't necessarily mean illness. A toxicologist and friend of mine, Dr. Emily Monosson, notes that the human body evolved to detoxify itself, to a point, yet new man-made chemical compounds weren't in the mix during this aeons-long process.

Despite growing evidence that human health has already been damaged by environmental chemicals, endocrine disruption theory has lurked at the fringes of mainstream medicine. It's been largely absent in medical school curricula and is rarely factored into medical diagnostics. But in September 2008, no less than the über-mainstream *Journal of the American Medical Association* published a study that may have marked a turning of the tide. The authors mined data from the CDC's 2002–2003 health survey, culling 1,455 subjects who tested positive for urinary BPA. They found that high BPA levels strongly, even convincingly, correlated with cardiovascular disease, diabetes, and liver enzyme abnormalities, all hallmarks of the modern health scourge, metabolic disorder.

Given industry claims that BPA safely whisks through the human body, you'd think some of us, some of the time, would be BPA-free. Since this is decidedly not the case, we must consider not only that BPA is hiding in every nook and cranny of our lives, but also that it actually lingers in our bodies longer than advertised. Frederick vom Saal, the controversial dean of BPA research, explained the seeming incongruity in a February 2011 podcast presented by *Environmental Health Perspectives*, the top journal in the field. Vom Saal said that the quick-metabolism model is a false industry claim meant to allay fears about the chemical's toxicity. His research has shown that BPA lingers in the system over days, not hours. As a result, he says, the average American chronically harbors eight times the federally set "safe" limit for BPA. He also noted that the rise of BPA in recent decades—production has more than doubled since the 1980s—tracks with the rise of obesity, type 2 diabetes, and cardiovascular disease. His rat studies show that BPA lowers sperm production and generates prostate cancer in males, and in females promotes early puberty, uterine fibroids, ovarian cysts, and miscarriage. These are human health trends. The dominant concern is that very small transplacental exposures in an unborn baby may reset its genetic coding, creating a person prone to certain diseases, disorders, and even strange behaviors, like those of Palanza's BPA-treated mice.

Phthalates, like BPA, are very well studied, with more than a thousand papers on record at PubMed. Among the most influential of these is University of Rochester researcher Shanna Swan's groundbreaking human study, published in 2005 by *Environmental Health Perspectives*. The study involved a cohort of 134 baby boys and their mothers. Maternal phthalate levels were measured before the women gave birth. When the boys were between two and thirty-six months old, Swan's team used calipers to measure the distance between the base of their penises and their anuses. A shorter distance indicates feminization. A "statistically significant" correlation was found between mothers with high phthalate levels and baby boys with feminized genitalia. Moreover, Swan extrapolated from the data that 25 percent of American mothers would harbor phthalate levels sufficient to feminize a son.

We may be closing in on the somewhat mysterious routes of BPA and phthalate contamination. A 2011 Berkeley study proved illuminating. A group of researchers selected five families on the basis of their high consumption of canned and packaged foods. The plan was to quantify the role that diet plays in the body's BPA and phthalate burdens. After initial baseline measurements were taken, the families were provided with fresh catered meals for three days. No processed, canned, or packaged foods or beverages were allowed. Daily urine samples were taken. The results were both clear and astonishing. BPA was reduced on average by more than 50 percent, and phthalates by more than 60 percent. And yet, in spring 2012, an FDA panel threw out a petition filed by the Natural Resources Defense

Council asking that BPA be banned from food packaging, claiming the petition lacked firm scientific proof of harm. Unsettled science . . .

In 2008, the U.S. Senate undertook a review of BPA. The investigating committee requested reports from several federal health agencies. The National Toxicology Program (NTP) submitted a 300-plus-page meta-study of BPA science to date. The NTP rates health risk from toxics on a "concern" scale. Certain aspects of BPA were accorded "some concern"—not high concern, but *some* concern . . . mid-range on a scale of one to five—primarily on the basis of rat studies. The studies had indicated possible effects on brain, behavior, and prostate glands in fetuses, infants, and children at current human exposure levels to BPA. That is, at exceedingly low parts-per-million (ppm) doses, equivalent to a single penny in a stack as tall as the Empire State Building. The brain effect noted was a reduction of "sexual dimorphism," meaning girls more like boys and vice versa, just what Palanza found in her BPA-dosed mice. (Palanza observed only behavior, however; she didn't dissect her subjects' brains as others later did.) The agency had *"minimal concern"* (emphasis added), about potential effects on breasts, early puberty in females, fetal or neonatal mortality, birth defects, and reduced birth weight and growth in newborns, despite vom Saal's findings of these very abnormalities in his rodent work. The result: a win for the chemical industry, which had funded science showing no harm.

Meanwhile, the EPA and the National Institute of Environmental Health Sciences have deployed tens of millions of Stimulus Act dollars to fund research designed to get to the bottom of BPA's effects on human health. One grant has gone to a project jointly proposed by the University of Illinois and Harvard. They will run side-by-side studies that track both rat and human newborns into early childhood, monitoring BPA exposures and checking for health problems. The focus will be on cognition and behavior in both study groups. A serious concern about EDCs such as BPA involves their suspected effects on the thyroid, the gland that governs fetal brain development and affects intellect.

An alarming new public health problem is the spike in autism rates. Once an extreme rarity, autism now strikes 1 in 54 boys. Philip Landrigan and other environmental health experts think environmental chemicals are likely autism triggers. Landrigan, a pediatrician and epidemiologist who chairs the Department of Preventive Medicine at Mount Sinai School of Medicine in New York, helped instigate the National Children's Study, now recruiting 150,000 as yet unborn subjects nationwide to be tracked for twenty-one years. The objective is to quantify everything about their lives, health, and environments and thereby obtain answers to a number of open questions about health and environment. Of special concern are the 3,300 high-production-volume (HPV) chemicals, produced at a rate of more than a million pounds a year and found in a wide array of consumer goods (including items made of plastic), cosmetics, medications, fuels, and building materials. A third of these common chemicals test positive for neurotoxicity in animals. Only eight have been scientifically proven neurotoxic to humans, since routine safety testing is not required for most of them. Landrigan counts BPA and brominated flame retardants among prime neurotoxic suspects, along with a number of solvents used in plastics production, and he fears that unborn babies are suffering lasting chemically induced brain impairments.

Could we rest easier about plastics safety if BPA and phthalates were banned? A 2010 study conducted at the University of

Texas suggests not. A team led by Dr. George Bittner tested 455 plastic products meant to be used with food or beverages. The items ranged from baby bottles and foam cups to plastic wrap and sandwich bags—some labeled as being BPA-free—from unnamed retail outlets, large, small, chain, and local. The goal was to measure estrogenic activity, if any, in these products without identifying particular chemicals. To this end, small squares of material were cut from each and subjected to physical stresses equivalent to repeated household dishwashing, freezing, microwaving, and exposure to UV rays in a closed car. Once properly abused, the samples were introduced to breast cancer cells, the "gold standard" for testing estrogenic activity. The result: Nearly every sample caused cell proliferation, including "BPA-free" baby bottles, plastic wrap, plant-based plastic, and "safe" plastics like high-density polyethylene and polypropylene. In fact, BPA-free baby bottles were more strongly estrogenic than those with BPA.

I e-mailed Bittner and asked if the result surprised him. He responded: "YES. Would have guessed 30%, not well over 90%." What Bittner and his colleagues found was that many under-the-radar chemicals associated with plastics have the potential to stimulate human estrogen receptors, the basic mechanism of endocrine disruption. Further, Bittner observed, "almost all of the five to thirty chemicals used to make a plastic part can leach from any plastic product because polymerization is almost always incomplete." Not only the residual feedstock monomers, the ethylene, propylene, and styrene, and so on, but also the "helper" chemicals, monomers that do not bind with plastic polymer chains, but catalyze, lend texture, color, feel, durability, and UV resistance; lubricate processing equipment; and prevent oxidation during processing. They all leach out into whatever comes into contact with the plastic. Bittner singles out phenols, not just bis*phenol*-A, as front-line estrogenic culprits. His study notes that a wide variety of phenols are used in "almost every phase" of plastics production. Estradiol, natural estrogen, is a phenol. So are the neurotransmitters serotonin, dopamine, and adrenaline. Consider the implications.

And there's more. Bittner's research team found that some materials that tested estrogen-free before manufacturing *became* estrogenic after processing and regular use. Bittner reported that normal stresses such as dishwasher, microwave, and UV can "change chemical structures or create chemical reactions to convert an estrogen-free chemical into a chemical having estrogenic action." On learning this, my thoughts turn to the millions of pounds of stressed-out plastics populating the oceans. . . .

Bittner points out that complex plastic products may harbor as many as one hundred different chemicals. Products like baby bottles, which have bottle, flange, and nipple components, all made from different plastic types. But as always, trade-secret laws protect the makers from having to divulge these synthetic ingredients, as Bittner ruefully notes. For those who believe consumer choice can force bad products off shelves, consider that consumers are often purposely kept in the dark about the true contents of the products they purchase. Until labeling requires disclosure of the ingredients in the package, not just the contents, the darkness will prevail.

Chemical compounds devised by humans in the past century are estimated to number about 100,000. Some 80,000 are in commercial use, and "thousands more [are] being added into commercial applications each year," according to an EPA newsletter. A small fraction, in the hundreds, has been thoroughly safety-tested by government agencies. About a thousand new compounds are registered with the EPA each year. Companies are required to provide chemical formulas and toxicity data, but only a small number of new chemical products come under direct scientific review by the EPA or the Food and Drug Administration. Meanwhile, agencies within the National Institutes of Health fund research into environmental contaminants and the CDC targets those 212 chemicals for the body-burden database. The research agencies can't enforce, the enforcement agencies do little research, and the taxpayer has a right to feel somewhat discomfited.

Here's how the American Chemistry Council—the deep-pocketed lobbying group for Dow, DuPont, and others—describes what *it* does: "The American Chemistry Council represents the companies that make the products that make modern life possible, while working to protect the environment, public health, and the security of our nation." Huh? Chemical companies relentlessly warn that increased regulation means job loss, stifled innovation, and a sinking Dow Jones. It's a message that strikes fear into the hearts of policy makers. But inaction makes us all guinea pigs in a massive but very profitable uncontrolled experiment that ignores the precautionary principle. Industry money quietly underwrites friendly research at prestigious institutions, including Harvard. Vom Saal studied BPA research outcomes according to funding sources. Nearly every independent study showed harm. Not one industry-funded study did. The point of most industry research is to *not* replicate adverse results. Each side accuses the other of bias.

The Toxic Substances Control Act (TSCA) was designed to protect the public from "unreasonable risk of injury to health or the environment" by regulating the manufacture and sale of chemicals, but it lacks enforcement teeth. By the time TSCA became law in 1976, problems from chemical company under-regulation were already glaring and included harm caused by DDT and PCBs. Leaded gasoline and house paints were still in use, though widely known to cause anemia and mental retardation, but it took until 1996 to phase them out, over industry protests. Only a few chemicals have been banned or restricted under the TSCA, and the 62,000 chemicals in use before its 1976 passage were "grandfathered." For a chemical to come under TSCA scrutiny, it has to pose "unreasonable risk" to health and environment. For the chemical lobby, that pretty much means dead bodies. The burden to prove harm rests with the EPA, an underfunded, understaffed, and beleaguered agency. It doesn't happen.

An effort to reform and update the TSCA came in 2010 in the form of the proposed Safe Chemicals Act. This law would require chemical producers to certify their products as safe *before* they can be marketed. It also calls for disclosure of product ingredients, much like food labeling, and for support for innovation in safer chemicals and products. In other words, it's designed to stimulate the economy while creating a healthier populace and planet. But the proposed law is not progressing in today's do-nothing Congress. Which is fine by the ACC, which calls the bill "burdensome" and of course job-killing. That hasn't stopped the State of California from passing ACC-opposed bills restricting the use of known toxics in products sold there.

Innovative technology may begin to sort out the toxics mess. A new program jointly operated by the NIH, EPA, and FDA, dubbed Tox21, will run chemicals through an ultra-high-tech robot programmed to screen for all manner of toxicity. The point is to replace laborious, expensive "observational" lab testing on animal subjects with an unbiased machine. This "high-through-put screening" will allow simultaneous testing of "many thousands of chemicals" in a matter of days. Ten thousand are queued up for analysis, and the National Toxicology Program is currently accepting nominations for additional chemical subjects. It's a new paradigm for the twenty-first century, reports the NTP website. Sounds good to me. Let's take out the guesswork and substitute standardized, comprehensive assays that might, as the NTP claims, be reliably predictive of health effects in people, not just rats. It's a system, I hope, that would be immune to industry charges of "junk science." Because, in a similar, unintended way, our bodies have been polluted much as the oceans have been.

ACKNOWLEDGMENTS

The authors have many to thank. First, we hope the fine people who appear in the book as key and supporting figures will hereby consider themselves thanked, and also honored for their essential work. The tireless efforts of our agent, Sandra Dijkstra, truly launched this book. We are also grateful to Taryn Fagerness and to Sandy's staff, most especially the preternaturally even-keeled Elise Capron. We have deeply appreciated and strongly benefited from the editorial acuity and philosophical commitment of our editor, Megan Newman. Dr. Emily Monasson was pivotal during the book's gestation and remained "on call." Similar gratitude is owed to Dr. Sarah Mosko and to the Algalita Marine Research Foundation staff, Marieta Francis and Jeanne Gallagher, and its chairman, Bill Francis.

Others who helped during the writing of this book include Drs. Jason Adolf, Karla McDermid, Hank Carson, Elizabeth Glover, and Judith Fox-Goldstein, all of the University of

Hawai'i at Hilo; Bonnie Monteleone of the University of North Carolina and Dr. Bill Cooper of the University of California, Irvine; Dr. Henrik Leffers of Copenhagen University; Dr. Rob Williams of the University of British Columbia; Dr. Elizabeth Venrick of the Scripps Institution of Oceanography; Dr. George Bittner of the University of Texas; Dr. David Caron of the Caron Lab, University of Southern California; Megan Lamson, of the Hawaii Wildlife Fund; Pete Leary of the U.S. Fish and Wildlife Service; Hayden Nevill of the International Bird Rescue's San Pedro Wildlife Care Center; Rick Anthony, cofounder of the California Resource Recovery Association and an early leader of the Zero Waste movement; and Bronwen Scott, developer of the Regional Reliance Inventory, which quantifies a region's local production capacity.

Charles wishes to especially thank Samala Cannon—my love of forty years, who keeps the sheets warm and the homestead habitable while I'm away at sea—and my parents, who while providing a lively intellectual environment allowed me to go my own way; John Herndon, my best friend in college, who provided critical thinking and made sure "my own way" wasn't ridiculous; my grandfather Will J. Reid, president of Hancock Oil, also a conservationist and the first president of Ducks Unlimited, and his foundation that helped fund the Algalita Marine Research Foundation. Barbara Fischer, who helped the Algalita Marine Research Foundation become a professional organization, as did Jim Ackerman, Nikhil Dave, Bill Grafton, and Marieta Francis. On the research side, thanks are due to Tony Andrady; Richard Thompson; Fred vom Saal; Jan Andres van Franeker; Charles Sheppard, editor of the *Marine Pollution Bulletin*; Bill Henry, who tracks Laysan albatross; and Cecilia Eriksson, who typed the plastic from our first research voyage to the gyre. Not only is "activist" not a pejorative term, it is what today's scientists must become if their work is to create a world in which life is really worth living. Thanks go to those who are active in their chosen fields: Jesse Goosens, author of *Plastic Soup*, Maria Westerbos, David Cooper, Vincent Petrus Janssen Steenburg, and Jan Andries von Franeker, who opened the door to the Netherlands; Myriam Zech, who cited our work when she wrote in the Los Angeles River TMDL for trash; Miriam Gordon, California Coastal Commission rep on our Prop 13 study and organizer of the Plastic Debris Rivers to Sea conference; Michael Bailey, videographer extraordinaire based in Hawaii; Eben Schwartz and Chris Parry of the California Coastal Commission Public Education Program, who saw the importance of the plastic pollution issue when I first told them; Paul Goettlich, who got an early head start on plastic toxicity with his website, www.mindfully.org; Stephanie Barger, CEO of Earth Resource Foundation and developer of the Campaign Against the Plastic Plague, so named by Jan Lundberg; and all those who have invited me to speak around the world. Thanks are due to my colleagues in Baja California: Laura Martinez of Pro Esteros, Gustavo Riano of Biopesca, and colleagues at the Autonomous University of Baja California, CICESE (Centro de Investigación Científica y de Educación Superior de Ensenada), and at the naval base, where a copy of *Synthetic Sea* in Español sits on their library shelf. Oceanographic research vessel *Alguita* has a lot of people to thank: Brett Crowther and Toby Richardson, who helped me change the original plans and build her at Richardson Devine Marine Constructions, then located on Macquarie Wharf in Hobart, Tasmania; Facundo Resendiz, my skilled first mate; the Southern California Marine Institute and Ken Kivett, captain of the *Sea Watch*, who showed me the ins and outs of research vessels and allowed me to borrow equipment; Alan Blunt at Seatek, who designed and installed *Alguita*'s replacement mast; Mike Benedict and Bill Campbell at Driscoll Boat Works in San Diego; Brian Scoles, who did a great job solarizing the vessel; Tomas Rojas, hydraulics engineer; and all the other tradespeople and volunteers who keep *Alguita* afloat and ready to sail off into the gyre.

CASSANDRA wishes to commend the USDA Small Business Innovation and Research Program, a productive yet endangered federal grant program that enables out-of-the-way entrepreneurs to develop and commercialize their ideas. In the strange interconnecting way of things, it led me to *Plastic Ocean*. Here's to you, Sandy, for your steadfast support over many years. To my husband, Bob Burkey, and my children, Billy and Keely: you are everything to me. Finally, I wish, with love, to dedicate my work on this book to my remarkable parents, James A. Phillips and Lauris Jardine Phillips.

SOURCES AND RESOURCES

Not all of these sources are cited in *Plastic Ocean*, but each helped shape it. The reader is encouraged to dive deeper into the issues, using them as resources.

BOOKS

Cradle to Cradle: Remaking the Way We Make Things by William McDonough and Michael Braungart. North Point Press, 2002.

Doubt Is Their Product: How Industry's Assault on Science Threatens Your Health by David Michaels. Oxford University Press, 2008.

Empires of Ice: Encounters in a Changing Landscape by Gretel Ehrlich. National Geographic, 2010.

The End of the Long Summer: Why We Must Remake Our Civilization to Survive on a Volatile Earth by Dianne Dumanoski. Crown Publishers, 2009.

Eye of the Albatross: Visions of Hope and Survival by Carl Safina. Owl Books, 2002.

Flotsametrics and the Floating World: How One Man's Obsession with Runaway Sneakers and Rubber Ducks Revolutionized Ocean Science by Curtis Ebbesmeyer and Eric Scigliano. Smithsonian Books/HarperCollins Publishers, 2009.

Garbage Land: On the Secret Trail of Trash by Elizabeth Royte. Back Bay Books/Little, Brown and Company, 2005.

Gone Tomorrow: The Hidden Life of Garbage by Heather Rogers. The New Press, 2005.

Humanity on a Tightrope: Thoughts on Empathy, Family and Big Changes for a Viable Future by Paul R. Ehrlich and Robert E. Ornstein. Rowman and Littlefield Publishers, 2010.

Industrial Plastics: Theory and Applications, third edition, by Terry L. Richardson, Ph.D., and Erik Lokensgard. Delmar Publishers, 1996.

Living Downstream: An Ecologist's Personal Investigation of Cancer and the Environment, second edition, by Sandra Steingraber. Da Capo Press, 2010.

Managing Without Growth: Slower by Design, Not Disaster by Peter A. Victor. Edward Elgar Publishing, 2009.

Marine Debris: Sources, Impacts and Solutions edited by James Coe and Donald Rogers. Springer-Verlag, 1997.

One-Dimensional Man by Herbert Marcuse. Beacon Press, 1966.

Our Stolen Future: Are We Threatening Our Fertility, Intelligence, and Surival? A Scientific Detective Story by Theo Colburn, Dianne Dumanoski, and John Peterson Myers. Plume/Penguin Group, 1996.

Plastics and the Environment edited by Anthony L. Andrady. Wiley and Sons, 2003.

The Polluters: The Making of Our Chemically Altered Environment by Benjamin Ross and Steven Amter. Oxford University Press, 2010.

Reason and Revolution: Hegel and the Rise of Social Theory by Herbert Marcuse. Beacon Press, 1970.

The Sea Around Us by Rachel Carson. Oxford University Press, 1951.

The Sea Can Wash Away All Evils: Modern Marine Pollution and the Ancient Cathartic Ocean by Kimberly C. Patton. Columbia University Press, 2007.

Sea Change: A Message of the Oceans by Sylvia A. Earle. G. P. Putnam's Sons, 1995.

Silent Spring by Rachel Carson. Houghton Mifflin Company, 1962.

The Waste Crisis: Landfills, Incinerators, and the Search for a Sustainable Future by Hans Tammemagi. Oxford University Press, 1999.

The Waste Makers by Vance Packard. David McKay Company, 1960.

Waste and Want: A Social History of Trash by Susan Strasser. Henry Holt, 1999.

PEER-REVIEWED SCIENTIFIC PAPERS

When I began studying marine debris, papers focusing on plastics were few and far between. Now they're among the most prolific.

"Biobased Performance Bioplastic: Mirel" by B. E. DeGregorio. *Chemistry and Biology Innovations* (2009). DOI 10.1016/j.chembiol.2009.01.001.

"A Comparison of Neustonic Plastic and Zooplankton Abundance in Southern California's Coastal Waters" by C. J. Moore, S. L. Moore, S. B. Weisberg, G. Lattin, and A. Zellers. *Marine Pollution Bulletin* 44 (2002): 1035–1038.

"A Comparison of Neustonic Plastic and Zooplanton at Different Depths Near the Southern California Shore" by G. L. Lattin, C. J. Moore, S. L. Moore, S. B. Weisberg, and A. Zellers. *Marine Pollution Bulletin* 49 (2004): 291–294.

"A Comparison of Plastic and Plankton in the North Pacific Central Gyre" by C. J. Moore, S. L. Moore, M. K. Leecaster, and S. B. Weisberg. *Marine Pollution Bulletin* 42 (2001): 1297–1300.

"Composition and Distribution of Beach Debris in Orange County, California" by S. L. Moore, D. Gregorio, M. Carreon, M. K. Leecaster, and S. B. Weisberg. *Marine Pollution Bulletin* 42 (2001): 241–245.

"Degradable Polyethylene: Fantasy or Reality" by K. R. Prasun, M. Hakkarainen, I. K. Varma, and A.-C. Albertsson. *Environmental Science Technology* 45, no. 10 (2011): 4217–4227.

"The Effects of Ingested Plastic on Seabirds: Correlations Between Plastic Load and Body Condition" by P. G. Ryan. *Environmental Pollution* 46 (1987): 119–125.

"Ingested Microscopic Plastic Translocates to the Circulatory System of the Mussel, *Mytilus Edulis*" by M. Browne, A. Dissanayake, T. Galloway, D. Lowe, and R. Thompson. *Environmental Science Technology* 42 (2008): 5026–5031.

"Lost at Sea: Where Is All the Plastic?" by R. Thompson, Y. Olsen, R. P. Mitchell, A. Davis, S. J. Rowland, A. W. G. John, G. McGonigle, and A. E. Russel. *Science* 304 (2004): 838.

"Marine Debris Collects Within the North Pacific Subtropical Convergence Zone" by W. G.

Pichel, J. Churnside, T. Veenstra, D. Foley, K. Friedman, R. Brainard, J. Nicoll, Q. Zheng, and
P. Clemente-Colón. *Marine Pollution Bulletin* 54 (2007): 1207–1211.

"Monitoring the Abundance of Plastic Debris in the Marine Environment" by P. G. Ryan, C. J.
Moore, J. A. van Franeker, and C. L. Moloney. *Philosophical Transactions of the Royal Society*
B 364 (2009): 1985–1998.

"Origins and Biological Accumulation of Small Plastic Particles in Fur Seal Scats from Macquarie Island" by C. Eriksson and H. Burton. *AMBIO* 32 (2003): 380–384.

"Patterns in the Abundance of Pelagic Plastic and Tar in the North Pacific Ocean, 1976–1985"
by R. H. Day and D. G. Shaw. *Marine Pollution Bulletin* 18 (1987): 311–316.

"Persistent Organic Pollutants Carried by Synthetic Polymers in the Ocean Environment" by
L. M. Rios, C. Moore, and P. R. Jones. *Marine Pollution Bulletin* 54 (2007): 1230–1237.

"Plastic Ingestion and PCBs in Seabirds: Is There a Relationship?" by P. G. Ryan, A. D. Connell,
and B. D. Gardner. *Marine Pollution Bulletin* 19 (1990): 174–176.

"Plastic Ingestion by Planktivorous Fishes in the North Pacific Central Gyre" by C. M. Boerger,
G. L. Lattin, S. L. Moore, and C. J. Moore. *Marine Pollution Bulletin* 60 (2010): 2275–2278.

"Plastic Resin Pellets as a Transport Medium for Toxic Chemicals in the Marine Environment" by
M. Yukie, I. Tomohiko, T. Hideshige, K. Haruyuki, O. Chiyoko, and K. Tsuguchika. *Environmental Science Technology* 35 (2001): 318–324.

"The Plastic World: Sources, Amounts, Ecological Impacts and Effects on Development, Repro-
duction, Brain and Behavior in Aquatic and Terrestrial Animals and Humans" by F. S. vom
Saal, S. Parmigiani, P. L. Palanza, L. G. Everett, and R. Ragaini. *Environmental Research* 108
(2008): 127–130.

"The Pollution of the Marine Environment by Plastic Debris: A Review" by José G. B. Derraik.
Marine Pollution Bulletin 44 (2002): 842–852.

"Quantification of Persistent Organic Pollutants Absorbed on Plastic Debris from the Northern
Pacific Gyre's 'Eastern Garbage Patch'" by L. Rios, P. Jones, C. Moore, and U. Narayan.
Journal of Environmental Monitoring 12 (2010): 2226–2236.

"Quantitative Analysis of Small Plastic Debris on Beaches in the Hawaiian Archipelago" by K.
J. McDermid and T. L. McMullen. *Marine Pollution Bulletin* 48 (2004): 790–794.

"Synthetic Polymers in the Marine Environment: A Rapidly Increasing, Long-Term Threat" by
C. J. Moore. *Environmental Research* 108 (2008): 131–139.

"Transport and Release of Chemicals from Plastics to the Environment and to Wildlife" by
E. Teuten, J. Saquing, D. Knappe, M. A. Barlaz, S. Jonsson, A. Bjorn, S. J. Rowland, R. C. Thomp-
son, T. S. Galloway, R. Ymashita, D. Ochi, Y. Waanuki, C. Moore, P. H. Viet, T. S. Tana,
M. Prudente, R. Boonyatumanond, M. P. Zakaria, K. Akkhavong, Y. Ogata, H. Hirai, S. Iwasa,
K. Mizukawa, Y. Hagino, A. Imamura, M. Saha, and H. Takada. *Philosophical Transactions of the
Royal Society* B 364 (2009): 2027–2045.

THE INTERNET

The Internet has transformed research into a trip down an almost bottomless rabbit hole. Here's
a selection of websites that proved indispensable or irresistible, grouped by category.

Stewardship and Monitoring

http://www.algalita.org/index.php. The website for the Algalita Marine Research Foundation.
Free access to published scientific papers typically "pay-per-view" at Internet science portals
like Elsevier. Also, voyage blogs, videos, unpublished research and conference proceedings,
and links to like-minded sites.

http://www.5gyres.org. Experts blog-tracking the plastic-sampling project unfolding in the
oceans' five gyres, a project spearheaded by Dr. Marcus Eriksen and Anna Cummins of Algalita. Valuable contextual information and links.

http://www.pelletwatch.org. Hideshige Takada's polluted pellet project, still open for business.

http://www.jean.jp/e_index.html. Japan Environmental Action Network, conducting marine litter investigations and clean-up activities.

http://chrisjordan.com/gallery/midway/#CF000313%2018x24. Chris Jordan focuses on the effects of mass consumption and production of goods through art and photography.

http://activities.cleanuptheworld.org/?581/. A UN program that mobilizes 35 million people in 130 countries each year "to clean up, fix up and conserve their environment."

http://marinemammalcenter.org. The premiere rehabilitation and research center for pinnipeds, based in the Bay Area. Heartwarming case histories, troubling studies.

http://meriresearch.org. The Maine-based nonprofit founded by Dr. Susan Shaw, the marine toxicologist who runs the Seals as Sentinels project, among other worthy efforts.

http://www.oceanconservancy.org. Annual International Coastal Cleanup data-set "snapshots" of shoreline pollution, underwritten by Coca-Cola, Dow, and Solo Cup.

http://www.surfrider.org. Surfers banded together to preserve beach access, then morphed into shoreline eco-activists. Its weekly e-newsletter, "Soup," is lively and smart.

http://www.imares.wur.nl/research/dossiers/plastic. Patch site that monitors fulmar ingestion of plastic.

Government Agencies and Publications

http://www.sccwrp.org/homepage.aspx. The Southern California Coastal Water Research Project offers exemplary marine science focusing on California coastal waters.

http://www.noaa.gov. National Oceanographic and Atmospheric Administration, managing polluted oceans, endangered marine species, and commercial fishing activities. Sign up for its e-newsletter: "Marine Debris Weekly."

http://www.pupuhanamokuakea.gov. All about the Northwest Hawaiian Islands, including agency and volunteer efforts to restore critical habitats and endangered wildlife.

http://www.epa.gov. Stats, status, and research pertaining to all things polluting. Also, information on their green chemistry initiatives and extended producer responsibility.

http://www.cdc.gov. Centers for Disease Control and Prevention. Portal to government programs aimed at keeping Americans healthy by learning what's making them sick.

http://www.niehs.nih.gov. National Institute of Environmental Health Sciences, repository of data about health risks posed by all manner of "environmental exposures."

http://www.ehp03.niehs.hih.gov/home.action. *Environmental Health Perspectives*, the top journal for studies and news about interactions between organisms and toxic agents.

http://www.ncbi.nlm.gov/pubmed. The National Center for Biotechnology Information, National Library of Medicine. Epic compendium of human health study abstracts.

http://memory.loc.gov/ammem/coolhtml/coolhome.html. A fascinating Library of Congress archive: *Prosperity and Thrift: The Coolidge Era and the Consumer Economy, 1921–1928.* Primary sources bear witness to the genesis of consumerism and disposability.

http://www.unep.org/. The United Nations Environmental Programme. Reports and data about the health of, and problems concerning, the oceans and other global ecosystems

Information Aggregators

http://www.sciencedaily.com. The latest science about everything. Bookmark it!

http://www.mindfully.org. Publications pulled in from everywhere dealing with the major environmental issues.

http://www.treehugger.com. Manages to be lively and truly green, despite having not-so-green advertisers, owing to Discovery Channel sponsorship.

http://www.outstolenfuture.org. The Internet extension of the book. Maintains an up-to-date catalog of studies that link hormone-disrupting agents and human health.

http://www.wikipedia.com. Comes in handy.

Nongovernmental Organizations/Nonprofits

http://earthjustice.org. "Because the earth needs a good lawyer." Sued the National Marine Fisheries Service on behalf of the Hawaiian monk seal. Forces enforcement.

http://www.environmentalhealthtrust.org. Founded by activist epidemiologist and author Devra Lee Davis, who seeks to make "prevention the cure."

http://www.ewg.org. The Environmental Working Group, a high-profile NGO with clout. Conducts research, lobbies, and maintains a highly informative website.

http://www.greensciencepolicy.org. Cofounded by Dr. Arlene Blum, biophysical chemist, extreme mountaineer, and leading anti-halogen flame-retardant activist.

http://www.plasticfreelife.com. News portal about plastic pollution.

http://plasticpollutioncoalition.org. Group started by Dianna Cohen and Daniella Russo now has branches around the world.

http://www.plasticpollutioncoalition.org. Santa Monica–based mélange of celebs, Ph.D.s, and just folks. Active, committed, worth Facebook-friending for regular posts and alerts.

http://www.psr.org. Physicians for Social Responsibility. With roots in nuke opposition, fights man-made threats to human health, including chemicals associated with plastics.

http://steadystate.org/. Center for the Advancement of the Steady State Economy. Bringing sanity to the subject of unsustainable economic growth.

Industry

http://www.americanchemistry.com. The American Chemistry Council, lobbyists for the chemical and plastics industries. Production stats and nice plastic information.

http://www.bicworld.com. Growth + green. How do they do it? Do they?

http://www.foodproductiondaily.com. Food and packaging industry scoop with an international perspective.

http://www.pubs.acs.org. Publications of the American Chemical Society. A highly rewarding site that offers both historical and new information about applied chemistry.

http://www.plasticsindustry.org. Website of the Society of Plastics Industries. For the trade. Home of Operation Clean Sweep, striving for a nurdle-free natural world.

Favorite Blogs

http://www.orvalguita.blogspot.com. The blog of oceanographic research vessel *Alguita* and her many scientific adventures.

http://www.boogiegreen.com. Dr. Sarah "Steve" Mosko brings sharp intelligence and keen wit to her research and writing about pollutants of the Plastic Age.

http://cleanbinproject.com/. A couple tries to buy nothing and create zero waste for a year.

http://www.myplasticfreelife.com. Formerly known as "fakeplasticfish.com," this blog chronicles the daily adventures of a woman learning how to rid everyday life of plastic.

http://www.theneighborhoodtoxicologist.blogspot.com. Dr. Emily Monasson deftly dispenses rooted-in-real-life information about toxicants in our midst.

http://www.peteatmidway.blogspot.com. Pete Leary, a wildlife biologist with the Fish & Wildlife Service, posts charming wildlife photos and stories about life on faraway Midway.

INDEX